SpringerBriefs in Molecular Science

Biobased Polymers

Series editor

Patrick Navard, Sophia Antipolis cedex, France

Published under the auspices of EPNOE*Springerbriefs in Biobased polymers covers all aspects of biobased polymer science, from the basis of this field starting from the living species in which they are synthetized (such as genetics, agronomy, plant biology) to the many applications they are used in (such as food, feed, engineering, construction, health, ...) through to isolation and characterization, biosynthesis, biodegradation, chemical modifications, physical, chemical, mechanical and structural characterizations or biomimetic applications. All biobased polymers in all application sectors are welcome, either those produced in living species (like polysaccharides, proteins, lignin, ...) or those that are rebuilt by chemists as in the case of many bioplastics.

Under the editorship of Patrick Navard and a panel of experts, the series will include contributions from many of the world's most authoritative biobased polymer scientists and professionals. Readers will gain an understanding of how given biobased polymers are made and what they can be used for. They will also be able to widen their knowledge and find new opportunities due to the multidisciplinary contributions.This series is aimed at advanced undergraduates, academic and industrial researchers and professionals studying or using biobased polymers. Each brief will bear a general introduction enabling any reader to understand its topic.

*EPNOE The European Polysaccharide Network of Excellence (www.epnoe.eu) is a research and education network connecting academic, research institutions and companies focusing on polysaccharides and polysaccharide-related research and business.

More information about this series at http://www.springer.com/series/15056

Soon Yee Liew · Wim Thielemans
Stefan Freunberger · Stefan Spirk

Polysaccharide Based Supercapacitors

 Springer

Soon Yee Liew
Division of Manufacturing and Process
   Technologies, Faculty of Engineering
University of Nottingham
Nottingham
UK

Wim Thielemans
Renewable Materials and Nanotechnology
   Research Group
KU Leuven
Kortrijk
Belgium

Stefan Freunberger
Institute for Chemistry and Technology
   of Materials
Graz University of Technology
Graz
Austria

Stefan Spirk
Institute for Chemistry and Technology
   of Materials
Graz University of Technology
Graz
Austria

and

Institute for the Engineering and Design
   of Materials
University of Maribor
Maribor
Slovenia

ISSN 2191-5407   ISSN 2191-5415   (electronic)
SpringerBriefs in Molecular Science
ISSN 2510-3407   ISSN 2510-3415   (electronic)
Biobased Polymers
ISBN 978-3-319-50753-8   ISBN 978-3-319-50754-5   (eBook)
DOI 10.1007/978-3-319-50754-5

Library of Congress Control Number: 2017933546

Printed on acid-free paper

This Springer imprint is published by Springer Nature
The registered company is Springer International Publishing AG
The registered company address is: Gewerbestrasse 11, 6330 Cham, Switzerland

Contents

Abstract

This work aims at providing a broad overview on the use of polysaccharides in supercapacitor applications, which is divided into three sections. The first section gives an introduction in the design and working function of supercapacitors in general as well as the very basics to understand the underlying electrochemistry. In the second section, composites of polysaccharide-based materials with conductive polymers are investigated and major achievements involving these in supercapacitor manufacturing are highlighted. The last section deals with so-called biocarbons, prepared by carbonization of polysaccharides either in pure form or as component of plant wastes.

Chapter 1
Introduction

Environmental concerns urge mankind to shift their energy supply from the currently mostly used non-renewable fossil ones towards sustainable and renewable sources. Apart from hydroelectricity most renewable energy sources, such as wind, solar, and tidal, are intermittent in nature and in general availability does not match demand. Consequently, efficient electrical energy storage will be a key enabler for a renewable energy supply. With regard to the decarbonization of traffic, which is a main source of greenhouse gas emissions, electrification is the key hurdle yet to be overcome. Electrochemical energy storage appears in either case a very desirable option; in the case of transport it might be even irreplaceable. The major storage principles involve rechargeable batteries and supercapacitors. Batteries have already revolutionized portable electronics and are currently entering the electric vehicle and stationary storage market. Similarly, supercapacitors appear as key component in many emerging applications including mobile devices, load shaving, and electrically operated trains. However, in order to make significant impact on a broad scale in an economic manner and with favorable ecological impact, yet significant improvements need to be achieved for both storage principles [1–3]. If electrochemical energy storage is to make maximum impact, it requires new generations with higher energy density, lower cost, longer lifetime and improved materials sustainability.

Firstly, we shall examine the working principle and characteristics of supercapacitors as opposed to batteries. Special attention will be given to different already used materials and their link to the storage mechanism. Consideration is also given to pseudo-capacitive materials and electrolytes. After considering applications of supercapacitors, we examine two major groups of bio based active materials in more details, namely composites of conducting polymers with cellulose nanomaterials and bio-derived carbon materials.

© The Author(s) 2017
S. Yee Liew et al., *Polysaccharide Based Supercapacitors*,
Biobased Polymers, DOI 10.1007/978-3-319-50754-5_1

1.1 Working Principle of a Supercapacitor

Batteries store electrical charge bulk solid redox materials that balance the electronic charge by the uptake or removal of counter ions into the material. This lends batteries superior energy density due to the high density of redox moieties in solid redox materials, albeit at rather low rate due to the need for ion diffusion in solids. In contrast, electrochemical double layer capacitors (EDLC) store charge by electrostatic sorption of ions at the electrode/electrolyte interface. A purely electrostatic EDLC consists of two porous electron conducting inert electrodes immersed in an ion-conducting and electron-insulating electrolyte (Fig. 1.1a). Upon polarization of one electrode against the other, the ions of the opposite sign will be drawn to the electrode and accumulate at the surface in an amount proportional to the electronic current flowing in the outer circuit. This forms the so-called electrical double layer or Helmholtz layer that consists of a space charge region in the vicinity of the interface; an electrical space charge zone at the electrode side and an ion space charge zone at the electrolyte side. The charge storage mechanism is purely electrostatic and non-Faradaic. This means that there is no charge transfer across the interface towards a species in the electrolyte or at the surface. Due to the absence of Faradaic reactions the electrodes are blocking and hence the rate is not limited by the activation overpotential as is the case with batteries. This allows for very fast rates, which are limited by Ohmic resistances only. Special attention has therefore to be taken with current collectors and cell construction in order to reduce these

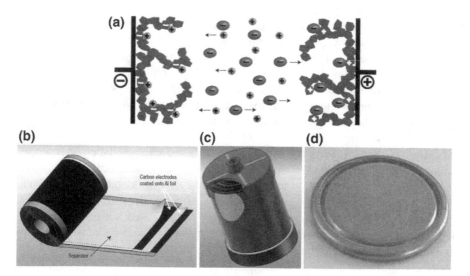

Fig. 1.1 a Principal setup of an EDLC with porous carbon electrodes on current collectors separated by an ion conducting electrolyte. **b–d** Construction of a spirally would EDLC, the assembled device with 2600 F in its housing and a flat 5 F coin device. Reproduced from Ref. [4] with permission from Nature Publishing Group

resistances and enable maximum rates. Cell constructions include spirally wound types, Fig. 1.1b, c, and flat ones, Fig. 1.1d.

A second group of electrochemical capacitors (ECs) additionally involves fast and reversible charge transfer to surface or near-surface species for charge storage. These are known as pseudo-capacitors or redox capacitors. At this point, it should be noted that there is some ambiguity in the literature of what pseudo-capacitive behavior is how the distinction to a battery type storage mechanism is defined. Particularly for devices, where power and capacity characteristics approach the typical range of the other storage device, very often these terms are used in a misleading way [5, 6]. We shall elaborate on this point later on.

What distinguishes conventional solid state and electrolytic capacitors from supercapacitors (sometimes also called ultracapacitors) is the capacity per gram of electrode material, which is by a factor of hundreds to thousands higher for the latter. One major factor for this behavior is a large specific surface area of the used electrode materials which could easily reach 1000–2000 $m^2\ g^{-1}$. Yet there is still a significant power and energy gap between EDLCs or pseudocapacitors and batteries. Figure 1.2 shows typical ranges of specific power and specific energy for capacitors and batteries in doubly logarithmic scale, the so-called Ragone plot.

The double layer capacitance C (in Farad, F) described by Helmholtz in 1853 is given by

$$C = \frac{\varepsilon_r \varepsilon_0 A}{d}$$

(1.1)

Fig. 1.2 Ragone plot with accessible regions of specific power and energy for batteries and electrochemical capacitors. Times associated with the *diagonal lines* are time constants obtained by dividing energy by power. Reproduced from Ref. [4] with permission from Nature Publishing Group

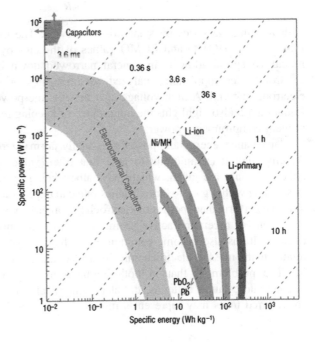

where ε_r and ε_0 are the dielectric constants of the electrolyte and vacuum, respectively, d is the effective thickness of the double layer and A the surface area. More recent models by Gouy and Chapman, and Stern and Geary include the description of a diffuse ion layer [7]. Depending on the used electrolyte, a capacitance of between 5 and 20 $\mu F\ cm^{-2}$ is found with the one in aqueous electrolytes being generally higher. Yet often organic electrolytes are used for their higher possible operating voltage that contributes with its value squared to the energy. The energy stored is determined by the capacitances of the negative and positive electrodes, C_- and C_+ and the maximum operating voltage V_{max} according to

$$E = \left(\frac{C_-C_+}{C_-+C_+}\right)V^2_{max} \tag{1.2}$$

With balanced electrodes it becomes

$$E = \frac{1}{2}C \cdot V^2_{max} \tag{1.3}$$

The stored charge Q (in coulombs, C, or mAh) at a given capacitance is

$$Q = C \cdot V \tag{1.4}$$

Being often the central reason for use the power is given by

$$P = \frac{V^2}{4R} \tag{1.5}$$

with R being the series resistance of the device. The maximum voltage is determined by the HOMO and LUMO values of the employed electrolyte. This theoretical voltage window is in general narrowed down by oxidative or reductive electrolyte decomposition catalyzed by impurities or functional groups of the electrode material. A high voltage also benefits the power with its value squared. Equation 1.5 also highlights the importance of keeping resistances small in order to prevent compromising power.

Capacitance, energy and power are usually normalized to the weight or volume. At this point it is important to note that often values per mass of the electron conductor only are given, which may, albeit being technically correct, be highly misleading with regard to "true" values one can expect on device level. This custom to relate performance metrics of electrochemical energy storage devices to the mass or volume of a certain "active" component is by no means restricted to supercapacitors. It has also become common in the field of batteries, where capacities are related to the mass of electrochemically active material or the electron conductor [2, 3]. The problem is that in both applications, supercapacitors and so called beyond-intercalation batteries, the electrochemical performance parameters which are referred to as may represent a minor to nearly negligible fraction of the total device mass or volume. This is even more problematic as publications often do not give all information to derive values for full electrode or device values including all other necessary components. The problem has been recognized for both superca-pacitors and batteries [8–10], best-practice publishing standards as recently pro-posed for solar cells are, however, yet to be defined [11]. For supercapacitor, the effect of different referencing is shown in Fig. 1.3. Energy and power are either referenced to mass and volume of carbon only or to carbon and current collector [8]. A material may show exceptional performance when values are referenced to the material alone, but inclusion of more cell components may curb the values massively. Particularly nanomaterials such as graphene have very low packing density. The empty space has to be filled with a large mass of electrolyte without adding capacity.

Strictly speaking, capacitive behavior requires the charge stored to depend lin-early on the potential change, i.e., capacitance is independent of potential. This is the case for purely electrostatic storage and gives rise to a rectangular shape of the current versus voltage curve. Pseudo-capacitance displays the same macroscopic behavior but may arise from completely different mechanism (involving Faradaic reactions). In a classical paper, Conway points out the thermodynamic differences between charge storage in capacitors and batteries and their experimental distinc-tion [12]. The key contents have recently been applied to classical and emerging materials by several authors [5, 6]. An illustrative summary is given in Fig. 1.4. Behavior of redox capacitors that feature more or less distinctive redox peaks in the current-voltage diagram has often equally been termed "pseudo-capacitive" although it does not suffice the strict definition [5, 6]. However, only where the forward and backward peaks in cyclic voltammogram are not separated there is reversible pseudocapacitance [6, 12]. Some of the most studied classes of materials that show pseudocapacitance are discussed in the next section.

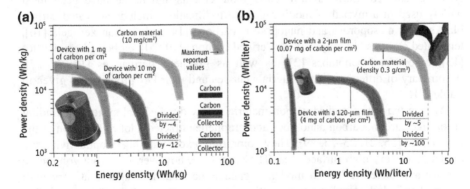

Fig. 1.3 Ragone plot for carbon materials with exceptional performance on a gravimetric (**a**) and volumetric (**b**) basis. Inclusion of more cell components, here the current collector, curbs values massively. Reproduced from Ref. [8] with permission from AAAS

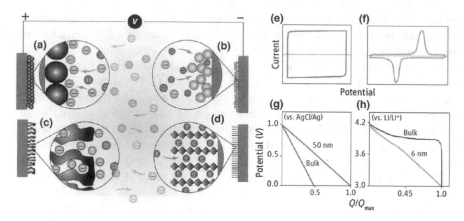

Fig. 1.4 a–d Charge storage mechanisms in capacitors. Electrostatic storage due to counter ion adsorption at carbon particles (**a**) or in pores (**b**). Pseudocapacitive redox capacitances at, e.g., RuO_2 (**c**) or intercalation pseudocapacitance with intercalation of Li+ ions (**d**). A capacitor responds with constant current to a linear potential change (**e**), which results linear voltage versus capacity dependence for both pure and pseudocapacitors (**g** shown for a bulk and nanoscale MnO_2 pseudocapacitor). In contrast show battery materials with Faradaic redox processes redox peaks (**f**). For nanoscale battery materials the V versus Q behavior approaches that of supercapacitors. Reproduced from Ref. [4] with permission from AAAS

1.2 Materials for Supercapacitors—High Surface Area Active Materials

A very high ion accessible surface area is a key parameter for achieving high double layer capacitance according to Eq. 1.1. At the same time electronic conductivity, electrochemical stability in a wide voltage window and ease of modification at low cost are required. Carbons fulfil these requirements and hence have been most widely used in a myriad of modifications. Modifications include activated carbon blacks [13], mesoporous and nanoporous carbons [14, 15], carbon xerogels [16], templated carbons [17–19], carbide derived carbons [19, 20], carbon fibers, fabrics [13], onions and nanotubes [21] as well as graphene derivatives [15, 22–27]. A summary of different major materials classes and their key properties is given in Table 1.1.

Among the many studied carbon materials, low dimensional systems including onion like carbon, carbon nanotubes and graphene attracted a lot of attention in the last decade [15, 21, 23–25, 29]. They commonly excel with high conductivity and moderate to very high surface area. They combine, however, only low volumetric density and therefore low to moderate capacitance that is available at very high power. Particularly graphene with its theoretical surface area of 2670 m^2/g suffers from restacking, which lowers the accessible surface area. Elaborate graphene architectures have therefore been proposed [24, 25]. A common feature of these material classes is their rather high cost. In contrast, the low cost of activated

Table 1.1 Carbon based materials for supercapacitors and their key properties

	Carbon onions	Carbon nanotubes	Graphene	Templated carbon	Carbide carbon	Activated carbon
Dimensionality	0D	1D	2D	3D	3D	3D
Surface area (m^2/g)	500–600	400–1200	≤ 2670	1000–3600	1000–3000	800–3500
Specific capacitance $(F g^{-1})$	≤ 30	≤ 70	100	80–200	70–180	50–300
Volumentric capacitance	Low	Low	Moderate	Low	High	High
Conductivity	High	High	High	Low	Moderate	Low
Cost	High	High	Moderate	High	Moderate	Low

Values given represent typical to upper values and refer to electrostatic storage in organic electrolyte. Table partially based on Refs. [3, 28]

carbons makes them the primary choice for most commercial devices. They are produced from organic precursors and subjected to an activation process. This activation process creates micropores (<2 nm) and mesopores (2–50 nm) concomitant with a very high accessible surface area that can exceed 2000 m^2/g. Activation processes include controlled oxidation with CO_2, water vapor, or strong bases. A very active field in activated carbons represents the use of a wide variety of natural precursors sourced from plants or animals. Heteroatoms such as O and N produce functional groups at the surfaces that reduce electric conductivity and require using additional conductive additives. Carbide derived carbons (CDC) represent a rather recent field in EC research [19, 20, 30]. They are based on leaching metal atoms from carbides, which produces micro and mesoporous carbons with controllable and tunable pore size distribution [30]. Nanopore size can be tuned between 0.6 and 1.1 nm. Another option to make controlled microporous carbon structures are inorganic [17–19] or soft organic templates [31] that allow 1, 2, and 3D structures to be produced.

Charge storage mechanism in pores. The traditional view on the origin of double layer capacitance is based on the charge separation in the Helmholtz layer, which involves the solvation sheath of the respective ion. Also more recent diffusive layer models involve solvation. Particularly in conjunction with the advent of CDCs, which allowed tuning of sub nanometer pores, it became, however, apparent that this picture fails to describe the behavior in very small pores. Using TiC derived CDCs Chmiola et al. found that below a size of 1.1 nm with Et_4N BF_4 (TEABF$_4$) in acetonitrile electrolyte the specific capacitance normalized to pore surface area increased sharply [32]. Such pores are smaller than the Helmholtz layer thickness. The effect could be explained by the partial desolvation of the ions entering the pore, which culminates in the formation of a wire-in-cylinder model. Huang et al. could closely fit the observed behavior by modelling solvated and desolvated ions in large and small pores, respectively [33]. The experimental and modelling results are

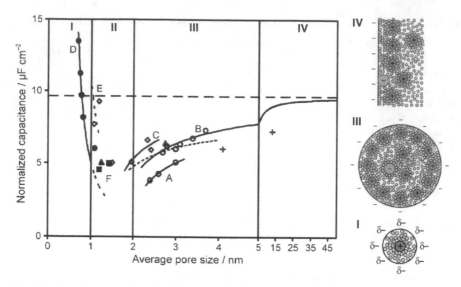

Fig. 1.5 Specific capacitance normalized to the BET surface area as a function of average pore diameter obtained with different carbon samples and the same 1.5 M Et_4NBF_4 in acetonitrile electrolyte (*filled* symbols, *open* symbols represent data with diverting salt concentration). *Full lines* represent model fits. Reproduced from Ref. [33] and with permission from Wiley

shown in Fig. 1.5. Small pores get increasingly desolvated (zone I), after the transition zone II the pore size is sufficiently large to accommodate a Helmholtz layer (zone III). With further increasing pore size the areal capacitance approaches that of a flat electrode (zone IV). Yet enhancing effects with reduced pore size hit limitations when pores become smaller than the naked ions or when they become very tortuous [34]. The finding of increasing capacity with confined ions spurred tremendous research efforts ever since, both in experiment and theory [32, 33, 35–43]. Recent efforts go therefore towards designing materials optimal for the respective ions at positive and negative electrode and for particular electroyltes [28, 41].

Pseudo-capacitive charge storage materials. Several classes of fast, reversible redox reactions at the surface of active materials exhibit pseudocapacitive behavior according to the discussion above. These include metal oxides [44] including MnO_2, Fe_3O_4 and RuO_2, carbon materials with surface functional groups and electronically conductive polymers [45]. Involving redox reactions results in typically lower cycling stability of the device than EDCLs. With the mentioned oxides the pseudo-capacitive storage involves surface adsorption of alkaline metal cations and insertion of protons. Nanosizing active materials have turned electrode materials for batteries that were previously considered too insulating into viable materials for even high power applications [46–48]. Along the same line, researchers have worked on improving the rate capability and capacitance of pseudocapacitive materials [27, 49]. Considering that pseudocapacitve materials store the ions in the first few nanometer the capacitance should be maximized by limiting the materials dimensions to this size. Capacitances in the range of 1300 F g^{-1} have been achieved using nanosized materials such as vanadium nitride, MnO_2, Nb_2O_5, RuO_2, TiO_2, or V_2O [27, 49–51]. Conducting polymers have been investigated in a wide variety for ECs. Materials include polyaniline, polypyrrole, polythiophene and radical polymers [45, 52, 53]. Stability is, however, an issue with these materials particularly when used in bulk.

Electrolytes. The importance of a wide stability window for energy and power has been pointed out above. Therefore the switch from aqueous electrolytes with a practical value of around 0.9 V towards organic electrolytes with around 2.7 V represented a major boost for capacitance. Aqueous electrolytes comprise typically alkaline chlorides and nitrates and trimethylammonium nitrate as well as sulfuric acid and alkaline hydroxides. Concentrations range typically from one to several molar. With organic electrolytes acetonitrile (AN) and propylene carbonate (PC) are the standard solvents. Salts used include alkaline and tertiary alkylammonium cations and mostly BF_4^- anions. To a smaller extent other anions known from Li-ion battery electrolytes such as ClO_4^-, PF_6^- and TFSI$^-$ are used. AN electrolytes excel for high power ECs for their high conductivity reaching 60 mS cm^{-1}. For safety and toxicity reasons, it can be replaced by PC albeit the power drops due to the lower conductivity. To further expand the usable voltage window ionic liquids (IL) are increasingly investigated as electrolytes. They offer furthermore the advantage of much improved safety due to non-volatility and inflammability [54]. Their high viscosity and low conductivity at room temperature is, however, an obstacle for use in ambient conditions, which requires careful choice of the constituting ions [55]. It must be emphasized that ILs are not just another class of solvents, where otherwise the picture of ions solvated in the solvent sheath prevails. As long as they are solvent free they are ion-only media that are far from conventional diluted electrolytes. In particular the structure of the electrical double layer at electrified interfaces is dramatically different, which of course severely impacts the charge storage when used in EDLCs. Its full understanding is a very active topic of modern fundamental physics. The most recent up-to-date account of the current understanding of ILs at electrified interfaces was published by Fedorov and Kornyshev [7].

In all the electrode material examples shown above, whether used for pure electrostatic storage or pseudocapacitance, capacitance is related to and limited by accessible surface. This limitation can in principle be surmounted by involving fast redox reactions with electrolyte borne redox species. This was recently reported for aqueous electrolytes with, e.g., iodide, hydroquinone, $VOSO_4$ or p-phenylenediamine [56–61]. These works have shown that dissolved redox species can react with the same fast kinetics as electrostatic surface storage. However, the capacitance could only be slightly enhanced and self-discharge was prohibitively high because of the low solubility and high ion diffusivity [62]. More recent efforts with redox electrolyte go towards ionic liquid electrolytes with a redox moiety attached to an ion [63–66].

1.3   Applications of Supercapacitors

ECs can store and deliver energy at very high specific power. Compared to other electrochemical energy storage devices both power and cycle life is exceptional. As such they are predestined for applications where very high power and repetitions cannot be met by, e.g., batteries. Spurred by the rise of high power batteries their manufacturers argue that the products could replace ECs with the advantage of higher energy. However, due to the different charge storage mechanism, the time constants for best use are different (see Fig. 1.2). However, high power batteries can be configured to deliver peak power, but their mass, volume and cost will be higher than with the same power delivered by ECs. Additionally, high rate operation will lead to faster aging of the battery. Equally, ECs are configured to store and deliver larger amounts of energy over longer timescales. There have been recent reviews on applications of ECs [4, 67–69]. For small devices like cameras and mobile phones, ECs with a few Farad capacitance are used to buffer power for memory backup. There are cordless tools available that use ECs with a few tens of farads and that can power the device for some minutes before being recharged within minutes. An application that highlights the maturity of ECs with regard to performance, reliability and safety is the use of large modules of 100 F supercapacitors for the emergency door opening in the Airbus A380 jet. ECs are also commonplace in municipal busses to operate doors.

The most important market for ECs is direct drive energy storage in transportation in addition to the wide use in auxiliaries. The market includes hybrid vehicles, metros, and trams. They are both used to store breaking energy for subsequent acceleration and to power the vehicle over short distances between frequent stops. In hybrid-electric vehicles ECs are used in combination with batteries to recuperate breaking energy and to take peak power from the battery to improve its cycle life. The extent of use of ECs for these applications is, as with batteries, in part hampered by the high price and a lack of technical maturity. The price and performance of the carbon materials ECs hold therefore key to wider market penetration and warrant therefore the continued research effort, including those described in the following.

References

1. Larcher, D., Tarascon, J.M.: Towards greener and more sustainable batteries for electrical energy storage. Nat. Chem. **9**, 19 (2014)
2. Bruce, P.G., Freunberger, S.A., Hardwick, L.J., Tarascon, J.-M.: Li-O_2 and Li-S batteries with high energy storage. Nat. Mater. **11**, 19 (2012)
3. Choi, N.-S., Chen, Z., Freunberger, S.A., Ji, X., Sun, Y.-K., Amine, K., Yushin, G., Nazar, L. F., Cho, J., Bruce, P.G.: Challenges facing lithium batteries and electrical double-layer capacitors. Angew. Chem. Int. Ed. **51**, 9994 (2012)
4. Simon, P., Gogotsi, Y.: Materials for electrochemical capacitors. Nat. Mater. **7**, 845 (2008)

5. Brousse, T., Bélanger, D., Long, J.W.: To be or not to be pseudocapacitive? J. Electrochem. Soc. **162**, A5185 (2015)
6. Simon, P., Gogotsi, Y., Dunn, B.: Where do batteries end and supercapacitors begin? Science **343**, 1210 (2014)
7. Fedorov, M.V., Kornyshev, A.A.: Ionic liquids at electrified interfaces. Chem. Rev. **114**, 2978 (2014)
8. Gogotsi, Y., Simon, P.: True performance metrics in electrochemical energy storage. Science **334**, 917 (2011)
9. Freunberger, S.A.: The aprotic lithium-air battery: new insights into materials and reactions. In: *ECS Conference on Electrochemical Energy Conversion & Storage*. The Electrochemical Society, Glasgow (2015)
10. Obrovac, M.N., Chevrier, V.L.: Alloy negative electrodes for li-ion batteries. Chem. Rev. **114**, 11444 (2014)
11. A checklist for photovoltaic research. Nat. Mater. **14**, 1073 (2015). http://www.nature.com/nmat/journal/v14/n11/full/nmat4473.html
12. Conway, B.E.: Transition from "supercapacitor" to "battery" behavior in electrochemical energy storage. J. Electrochem. Soc. **138**, 1539 (1991)
13. Bao, L., Li, X.: Towards textile energy storage from cotton T-shirts. Adv. Mater. **24**, 3246 (2012)
14. He, X., Li, R., Qiu, J., Xie, K., Ling, P., Yu, M., Zhang, X., Zheng, M.: Synthesis of mesoporous carbons for supercapacitors from coal tar pitch by coupling microwave-assisted KOH activation with a MgO template. Carbon **50**, 4911 (2012)
15. Jiang, H., Lee, P.S., Li, C.Z.: 3D carbon based nanostructures for advanced supercapacitors. Energy Environ. Sci. **6**, 41 (2013)
16. Frackowiak, E., Beguin, F.: Carbon materials for the electrochemical storage of energy in capacitors. Carbon **39**, 937 (2001)
17. Vix-Guterl, C., Frackowiak, E., Jurewicz, K., Friebe, M., Parmentier, J., Beguin, F.: Electrochemical energy storage in ordered porous carbon materials. Carbon **43**, 1293 (2005)
18. Fuertes, A.B., Lota, G., Centeno, T.A., Frackowiak, E.: Templated mesoporous carbons for supercapacitor application. Electrochim. Acta **50**, 2799 (2005)
19. Korenblit, Y., Rose, M., Kockrick, E., Borchardt, L., Kvit, A., Kaskel, S., Yushin, G.: High-rate electrochemical capacitors based on ordered mesoporous silicon carbide-derived carbon. ACS Nano **4**, 1337 (2010)
20. Chmiola, J., Largeot, C., Taberna, P.L., Simon, P., Gogotsi, Y.: Monolithic carbide-derived carbon films for micro-supercapacitors. Science **328**, 480 (2010)
21. Pan, H., Li, J.Y., Feng, Y.P.: Carbon nanotubes for supercapacitor. Nanoscale Res. Lett. **5**, 654 (2010)
22. Zhai, Y.P., Dou, Y.Q., Zhao, D.Y., Fulvio, P.F., Mayes, R.T., Dai, S.: Carbon materials for chemical capacitive energy storage. Adv. Mater. **23**, 4828 (2011)
23. Wu, Z.S., Zhou, G.M., Yin, L.C., Ren, W., Li, F., Cheng, H.M.: Graphene/metal oxide composite electrode materials for energy storage. Nano Energy **1**, 107 (2012)
24. Huang, Y., Liang, J.J., Chen, Y.S.: An overview of the applications of graphene-based materials in supercapacitors. Small **8**, 1805 (2012)
25. Li, C., Shi, G.Q.: Three-dimensional graphene architectures. Nanoscale **4**, 5549 (2012)
26. Zhang, L.L., Zhao, X.S.: Carbon-based materials as supercapacitor electrodes. Chem. Soc. Rev. **38**, 2520 (2009)
27. Zhao, X., Sanchez, B.M., Dobson, P.J., Grant, P.S.: The role of nanomaterials in redox-based supercapacitors for next generation energy storage devices. Nanoscale **3**, 839 (2011)
28. Simon, P., Gogotsi, Y.: Capacitive energy storage in nanostructured carbon-electrolyte systems. Acc. Chem. Res. **46**, 1094 (2013)
29. Yu, H., Wu, J., Lin, J., Fan, L., Huang, M., Lin, Y., Li, Y., Yu, F., Qiu, Z.: A reversible redox strategy for SWCNT-based supercapacitors using a high-performance electrolyte. ChemPhysChem **14**, 394 (2013)

30. Leis, J., Arulepp, M., Kuura, A., Lätt, M., Lust, E.: Electrical double-layer characteristics of novel carbide-derived carbon materials. Carbon **44**, 2122 (2006)
31. Dipendu, S., Renju, Z., Amit, K.N.: Soft-templated mesoporous carbons: chemistry and structural characteristics. In: Polymer Precursor-Derived Carbon, vol. 1173, p. 61. American Chemical Society (2014)
32. Chmiola, J., Yushin, G., Gogotsi, Y., Portet, C., Simon, P., Taberna, P.L.: Anomalous increase in carbon capacitance at pore sizes less than 1 nanometer. Science **313**, 1760 (2006)
33. Huang, J., Sumpter, B.G., Meunier, V.: A universal model for nanoporous carbon supercapacitors applicable to diverse pore regimes, carbon materials, and electrolytes. Chem. Eur. J. **14**, 6614 (2008)
34. Kajdos, A., Kvit, A., Jones, F., Jagiello, J., Yushin, G.: Tailoring the pore alignment for rapid ion transport in microporous carbons. J. Am. Chem. Soc. **132**, 3252 (2010)
35. Largeot, C., Portet, C., Chmiola, J., Taberna, P.-L., Gogotsi, Y., Simon, P.: Relation between the ion size and pore size for an electric double-layer capacitor. J. Am. Chem. Soc. **130**, 2730 (2008)
36. Huang, J., Sumpter, B.G., Meunier, V.: Theoretical model for nanoporous carbon supercapacitors. Angew. Chem. Int. Ed. **47**, 520 (2008)
37. Merlet, C., Rotenberg, B., Madden, P.A., Taberna, P.-L., Simon, P., Gogotsi, Y., Salanne, M.: On the molecular origin of supercapacitance in nanoporous carbon electrodes. Nat. Mater. **11**, 306 (2012)
38. Limmer, D.T., Merlet, C., Salanne, M., Chandler, D., Madden, P.A., van Roij, R., Rotenberg, B.: Charge fluctuations in nanoscale capacitors. Phys. Rev. Lett. **111** (2013)
39. Merlet, C., Pean, C., Rotenberg, B., Madden, P.A., Daffos, B., Taberna, P.L., Simon, P., Salanne, M.: Highly confined ions store charge more efficiently in supercapacitors. Nat. Commun. **4** (2013)
40. Merlet, C., Rotenberg, B., Madden, P.A., Salanne, M.: Computer simulations of ionic liquids at electrochemical interfaces. Phys. Chem. Chem. Phys. **15**, 15781 (2013)
41. Pean, C., Merlet, C., Rotenberg, B., Madden, P.A., Taberna, P.-L., Daffos, B., Salanne, M., Simon, P.: On the dynamics of charging in nanoporous carbon-based supercapacitors. ACS Nano **8**, 1576 (2014)
42. Prehal, C., Weingarth, D., Perre, E., Lechner, R.T., Amenitsch, H., Paris, O., Presser, V.: Tracking the structural arrangement of ions in carbon supercapacitor nanopores using in situ small-angle X-ray scattering. Energy Environ. Sci. **8**, 1725 (2015)
43. Griffin, J.M., Forse, A.C., Tsai, W.-Y., Taberna, P.-L., Simon, P., Grey, C.P.: In situ NMR and electrochemical quartz crystal microbalance techniques reveal the structure of the electrical double layer in supercapacitors. Nat. Mater. **14**, 812 (2015)
44. Zhi, M.J., Xiang, C.C., Li, J.T., Li, M., Wu, N.Q.: Nanostructured carbon-metal oxide composite electrodes for supercapacitors: a review. Nanoscale **5**, 72 (2013)
45. Snook, G.A., Kao, P., Best, A.S.: Conducting-polymer based supercapacitor devices and electrodes. J. Power Sources **196**, 1 (2011)
46. Arico, A.S., Bruce, P.G., Scrosati, B., Tarascon, J.-M., van Schalkwijk, W.: Nanostructured materials for advanced energy conversion and storage devices. Nat. Mater. **4**, 366 (2005)
47. Bruce, P.G., Scrosati, B., Tarascon, J.-M.: Nanomaterials for rechargeable lithium batteries. Angew. Chem. Int. Ed. **47**, 2930 (2008)
48. Poizot, P., Laruelle, S., Grugeon, S., Dupont, L., Tarascon, J.M.: Nano-sized transition-metal oxides as negative-electrode materials for lithium-ion batteries. Nature **407**, 496 (2000)
49. Augustyn, V., Come, J., Lowe, M.A., Kim, J.W., Taberna, P.-L., Tolbert, S.H., Abruña, H.D., Simon, P., Dunn, B.: High-rate electrochemical energy storage through Li+ intercalation pseudocapacitance. Nat. Mater. **12**, 518 (2013)
50. Choi, D., Blomgren, G.E., Kumta, P.N.: Fast and reversible surface redox reaction in nanocrystalline vanadium nitride supercapacitors. Adv. Mater. **18**, 1178 (2006)

51. Brezesinski, K., Wang, J., Haetge, J., Reitz, C., Steinmueller, S.O., Tolbert, S.H., Smarsly, B. M., Dunn, B., Brezesinski, T.: Pseudocapacitive contributions to charge storage in highly ordered mesoporous group V transition metal oxides with iso-oriented layered nanocrystalline domains. J. Am. Chem. Soc. **132**, 6982 (2010)
52. Pan, L., Yu, G., Zhai, D., Lee, H.R., Zhao, W., Liu, N., Wang, H., Tee, B.C.-K., Shi, Y., Cui, Y., Bao, Z.: Hierarchical nanostructured conducting polymer hydrogel with high electrochemical activity. Proc. Natl. Acad. Sci. **109**, 9287 (2012)
53. Koshika, K., Sano, N., Oyaizu, K., Nishide, H.: An ultrafast chargeable polymer electrode based on the combination of nitroxide radical and aqueous electrolyte. Chem. Commun. 836 (2009)
54. Armand, M., Endres, F., MacFarlane, D.R., Ohno, H., Scrosati, B.: Ionic-liquid materials for the electrochemical challenges of the future. Nat. Mater. **8**, 621 (2009)
55. Lin, R., Taberna, P.-L., Fantini, S., Presser, V., Pérez, C.R., Malbosc, F., Rupesinghe, N.L., Teo, K.B.K., Gogotsi, Y., Simon, P.: Capacitive energy storage from −50 to 100 °C using an ionic liquid electrolyte. J. Phys. Chem. Lett. **2**, 2396 (2011)
56. Lota, G., Frackowiak, E.: Striking capacitance of carbon/iodide interface. Electrochem. Commun. **11**, 87 (2009)
57. Lota, G., Fic, K., Frackowiak, E.: Alkali metal iodide/carbon interface as a source of pseudocapacitance. Electrochem. Commun. **13**, 38
58. Senthilkumar, S.T., Selvan, R.K., Lee, Y.S., Melo, J.S.: Electric double layer capacitor and its improved specific capacitance using redox additive electrolyte. J. Mater. Chem. A **1**, 1086
59. Senthilkumar, S.T., Selvan, R.K., Melo, J.S.: Redox additive/active electrolytes: a novel approach to enhance the performance of supercapacitors. J. Mater. Chem. A **1**, 12386
60. Yu, H., Wu, J., Fan, L., Lin, Y., Xu, K., Tang, Z., Cheng, C., Tang, S., Lin, J., Huang, M., Lan, Z.: A novel redox-mediated gel polymer electrolyte for high-performance supercapacitor. J. Power Sources **198**, 402
61. Wang, Y., Cardona, C.M., Kaifer, A.E.: Molecular orientation effects on the rates of heterogeneous electron transfer of unsymmetric dendrimers. J. Am. Chem. Soc. **121**, 9756 (1999)
62. Sathyamoorthi, S., Suryanarayanan, V., Velayutham, D.: Organo-redox shuttle promoted protic ionic liquid electrolyte for supercapacitor. J. Power Sources **274**, 1135
63. Ghilane, J., Fontaine, O., Martin, P., Lacroix, J.C., Randriamahazaka, H.: Formation of negative oxidation states of platinum and gold in redox ionic liquid. electrochemical evidence. Electrochem. Commun. **10**, 1205 (2008)
64. Fontaine, O., Lagrost, C., Ghilane, J., Martin, P., Trippe, G., Fave, C., Lacroix, J.C., Hapiot, P., Randriamahazaka, H.N.: Mass transport and heterogeneous electron transfer of a ferrocene derivative in a room-temperature ionic liquid. J. Electroanal. Chem. **632**, 88 (2009)
65. Chen, X., Xu, D., Qiu, L., Li, S., Zhang, W., Yan, F.: Imidazolium functionalized TEMPO/iodide hybrid redox couple for highly efficient dye-sensitized solar cells. J. Mater. Chem. A **1**, 8759 (2013)
66. Yu, H., Fan, L., Wu, J., Lin, Y., Huang, M., Lin, J., Lan, Z.: Redox-active alkaline electrolyte for carbon-based supercapacitor with pseudocapacitive performance and excellent cyclability. Rsc Adv. **2**, 6736
67. Miller, J.R., Simon, P.: Electrochemical capacitors for energy management. Science **321**, 651 (2008)
68. Miller, J.R., Burke, A.F.: Electrochemical capacitors: challenges and opportunities for real-world applications. Electrochem. Soc. Interface **17**, 53 (2008)
69. Kötz, R., Carlen, M.: Principles and applications of electrochemical capacitors. Electrochim. Acta **45**, 2483 (2000)

Chapter 2
Polysaccharides in Supercapacitors

In this part, the use of polysaccharides, either directly through composite approaches, or by carbonization will be described. In many cases, materials are obtained which are competitive in terms of capacitance and cycle lifetime. In this part, the use of polysaccharides, either directly through composite approaches, or by carbonization will be described. In many cases, materials are obtained which are competitive in terms of capacitance and cycle lifetime. The following part will focus mainly on cellulosic composites with conductive polymers since cellulose is most abundant and therefore has attracted much more research interest in this field whereas in the second part also other polysaccharides, such as chitin, xylans, alginates, pectins, dextrans and caragenaans have been used in carbonization experiments.

2.1 Native Polysaccharides in Composites

The charge storage abilities of electronically conducting polymers (ECPs), which arise when they are repeatedly oxidized and reduced, make them very attractive materials for the fabrication of supercapacitor electrodes. This mechanism of charge storage is called pseudo-capacitance. Unlike double-layer capacitance that involves charging of only the surface of a material, pseudo-capacitance can take place throughout the volume of a material provided there is sufficient electrolyte access for the counter ions to accomplish the redox process. The theoretical mass specific capacitance of for example polyaniline (PANI) and polypyrrole (PPY) can be as high as 750 and 620 F g^{-1}, respectively [1]. This is many times that of the mass specific capacitance measured for typical carbon electrode materials and is the basis of the appeal of ECPs as potential supercapacitor electrode materials. One of major challenges facing the development of ECPs for practical use in supercapacitors is how to form a porous structured ECP electrode such that electrolyte transfers are unimpeded. This challenge is difficult to overcome if ECPs alone are used for the

© The Author(s) 2017
S. Yee Liew et al., *Polysaccharide Based Supercapacitors*,
Biobased Polymers, DOI 10.1007/978-3-319-50754-5_2

electrode fabrication. Indeed, ECPs are mechanically weak materials and do not favor the formation of extended porous structures. The other limitation posed by the weakness of ECP materials is the propensity of these electrode materials to degrade very quickly during repeated charge/discharge cycles; They are unable to withstand the swelling and shrinkage of the ECP when electrolyte transfers across the ECP/electrolyte interface. An interesting example to mention here is PANI. Owing to its molecular structure, PANI deposits are often very porous, which significantly favors the transport of electrolyte during the charge/discharge cycles. However, among the most common ECPs, PANI has the worst cycling stability, i.e., the capacitance degrades quickest per number of charge/discharge cycles when compared to other ECPs such as PPY and poly(3,4-ethylenedioxythiophene) (PEDOT), whose cycling stabilities are also poor. The poor cycling performance of PANI is a very good demonstration of the disadvantage of the inherent weakness of the polymer material. Despite the ease with which a porous PANI structure can be generated, it is the overall structure integrity, and therefore the cycling stability, that was worse than for non-porous PPY and PEDOT. One can thus conclude that although ECPs can demonstrate a very high electrochemical capacitance that makes them desirable for use in supercapacitors, practical applications of these materials limitations is limited due to their poor mechanical strength.

The mechanical weakness of ECPs during charge/discharge cycling can be easily overcome by forming nanocomposites of ECPs with rigid support materials such as carbon nanotubes (CNTs). This often results in mechanically stable nanocomposites [2–5]. Owing to the geometry of CNTs, their incorporation facilitates the formation of porous ECP nanocomposites as they become the template onto which the ECPs coat during the deposition process. CNTs will then form the backbone structure of the porous ECP-CNT nanocomposites, providing the mechanical integrity. The strength of the ECP-CNT nanocomposites also means that the porous structure can be extended to thicker deposit on electrodes. This allows the fabrication of high total capacitance electrodes, promoting the economical utilization of the geometric area of the collector towards practical supercapacitors for energy storage applications. For example, pellet electrodes had been made from chemically-synthesized PPY-CNT nanocomposites by pressing without any polymer binder and this electrode had a specific capacitance of 495 F g^{-1} [6]. The PPY-CNT nanocomposite was able to retain their porous internal structure when being pressed at 500 kg cm^{-2}, demonstrating the strength contributed by the MWCNTs [6]. By co-electrodeposition of PPY and CNTs, thick films of porous PPY-CNT could achieve an electrode specific capacitance of 2.35 F cm^{-2} [7]. Without CNTs, the electrodeposited PPY only reached around 0.7 F cm^{-2} [8]. More recently, a significant amount of similar research has been carried out on ECP nanocomposites with graphene, many of which showed promise for use in supercapacitors. However, as graphene is a sheet and therefore has a tendency to stack, it makes it much more difficult to create similar porous nanocomposite structures that are comparable to CNT-ECP structures.

Given that the role of CNTs in the production of stable ECP-CNT nanocomposites was mainly strength and the facilitation of the creation of a porous structure, it was thought that this role could also be satisfied with nanocelluloses such as cellulose nanocrystals (CNXLs) and cellulose nanofibrils (CNFs) instead of CNTs. In addition, the physical conditions (temperature, pressure and chemicals) under which the ECP-CNT nanocomposites are typically synthesized tend to be rather mild and could therefore also be applied to CNXLs and CNFs. For example, the reaction mixtures from which ECP nanocomposites are formed are mostly aqueous systems and not corrosive enough to cause damage to CNXLs or CNFs. CNXLs and CNFs display sufficient chemical resistance due to their crystalline structure (CNXLs are 88.6% [9] crystalline and above 70% crystallinity for CNFs [10, 11]), which makes that any possible degradation needs to degrade outer layers first before access to the inner chains is possible under non-swelling conditions. Perhaps of greater importance is that their rigid crystalline structures give rise to remarkable strength of the CNXLs and CNFs that can be used to form a mechanically robust template. The high aspect ratio of the CNXLs and the CNFs further means that they can also facilitate the formation of porous ECP nanocomposites similarly to CNTs. When compared to CNTs, nanocellulose materials are easier to prepare, safer to use as they are biocompatible and friendly to the environment, and significantly more cost effective [12–16]. Therefore, if ECP-nanocellulose nanocomposites can be made to perform equally well electrochemically as ECP-CNT nanocomposites, they could disrupt the prominence of ECP-CNT nanocomposites and open up more applications and markets for high-performing supercapacitors.

The ECP most commonly used for the fabrication of nanocomposites with CNXLs or CNFs is PPY. The main reason for this is the ease of processing of PPY; the PY monomer is soluble in water and therefore mixing of the pyrrole (PY) monomer with the supporting CNXLs and CNFs (which are stably dispersed in water) prior to the composite formation is straightforward. The other reason for the popularity of PPY use for nanocomposite making is due to its structure. When PPY is formed without the use of any supporting templates such as CNTs or CNXLs, the polymers have a non-porous, globular 'cauliflower' structure as a result of the α-β and β-β couplings that occur during polymerization (Fig. 2.1). This gives PPY by itself a relatively poor capacitive behavior during cycling. Therefore, when nanocomposites are formed using PPY together with supporting nanomaterials to form porous nanocomposites, structure comparison against the structure of pristine PPY can be done easily. This also makes it straightforward to relate the capacitive behavior improvement of the nanocomposites over pristine PPY as a result of their structural difference. The structural differences are also easily visualized and characterized using microscopy.

The formation of the ECP composite on the mechanically rigid nanoparticle template is usually carried out by either co-electrodeposition or by in situ chemical oxidation, where the monomer is polymerized in the presence of nanoparticles or the already assembled nanoparticle template either by an electrochemical oxidation or a chemical oxidation respectively. Compared to the co-electrodeposition method, an in situ chemical deposition offers the possibility of easier large-scale production, even though electrodeposition is widely used in industry for example for electroplating. In most cases, when a nanocomposite can be prepared by co-electrodeposition, which is

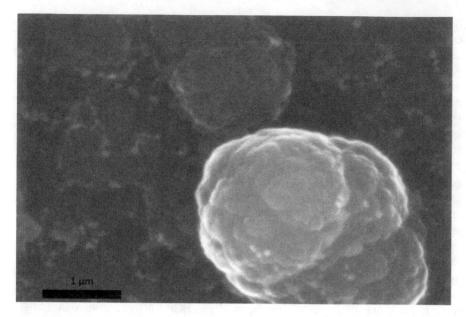

Fig. 2.1 Scanning electron microscopy image of electrodeposited PPY with chloride anions showing the typical 'cauliflower structure'

a more technically challenging process, it can also be produced by in situ chemical deposition methods. Therefore, it is important to investigate and compare the performance of the nanocomposites prepared by both methods.

Let's consider the implications of the two different preparation methods using the preparation of PPY/CNXL nanocomposites. During the co-electrodeposition process of PPY/CNXL, PPY is polymerized selectively and deposited when oxidative current passes at the working electrode by applying a positive potential. The formation of the PPY/CNXL nanocomposite requires the spontaneous transport of the negatively charged CNXLs towards the working electrode for the incorporation within the positively charged PPY matrix that has just been formed as the charge balancing species as well as the structural support [17]. This 'spontaneous' transport and incorporation of CNXLs into the nanocomposite requires (1) the active transport of individual negatively charged CNXLs toward the electrode surface by diffusion and (2) the strong electrostatic interaction between the CNXLs and the PPY that allows the PPY to wrap over the whole CNXL surface. During the in situ chemical deposition process, the oxidation of the monomers is carried out by the addition of oxidizing chemicals which will occur in the whole reaction volume. The polymerized ECP product is thus not confined to a specific area, and the strong electrostatic interaction between the CNXLs and the PPY that allows the PPY to wrap over the whole CNXL surface will thus also occur in the whole reaction volume. So while is it anticipated that coating of the CNXLs by PPY will occur,

Electrodeposited PPY/CNXL nanocomposite electrode film, showing a continuous ECP phase and good electrode contact

Electrode assembled from in-situ chemical deposited PPY/CNXL nanocomposite electrode film, showing a structure which consists of discontinuous PPY-coated CNXLs and poor electrode contact

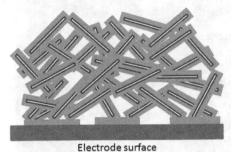
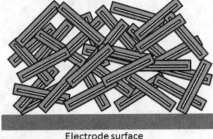

Electrode surface   Electrode surface

Fig. 2.2 Comparison schematic between (*left*) an electrodeposited PPY/CNXL nanocomposite electrode film and (*right*) an electrode assembled from the chemical deposited PPY/CNXL. *Solid lines* depict discrete interfaces

based on results from co-electrodeposition, there is thus no active driving force of the CNXLs towards the desired deposition surface. Therefore, the nanocomposite products of in situ deposition require assembly into an electrode for characterization and use. As a result, the solid products often suffer from high contact resistance due to the large number of interfaces that exist between the individual nanocomposite particulates. This problem is usually significantly less pronounced for the electrodeposited nanocomposite films as the nanocomposite is grown from the electrode directly and a continuous conducting pathway is ensured (Schematic representation in Fig. 2.2). This could be resolved by chemically polymerizing monomer into an existing porous template but care needs to be taken not to clog pores as they are needed for electrolyte transport.

2.2 ECP/CNXL Nanocomposites

The first PPY/CNXL nanocomposite films for supercapacitor applications have been fabricated by co-electrodeposition in 2010 [17]. Co-electrodeposition means that the CNXLs were incorporated simultaneously into the PPY during the electrodeposition as the counter anion species. The negative charge on the surface of the CNXLs was introduced by subjecting the CNXLs to a controlled oxidation process (mediated by TEMPO) [18]. This oxidation process converts the surface primary hydroxyl groups of the CNXLs to carboxylic acid groups, which deprotonate when the CNXLs are dispersed in water to give a negative charge on the CNXL surface. The co-electrodeposited PPY/CNXL films displayed a very porous structure (Fig. 2.3), comprising PPY-coated individual CNXLs, as opposed to the typical

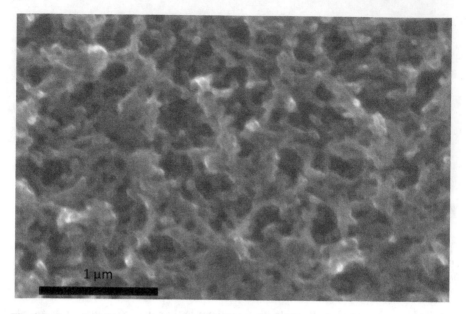

Fig. 2.3 Scanning electron micrograph of an electrochemically deposited PPY/CNXL nanocomposite film on glassy carbon electrode

globular structure of the PPY (Fig. 2.1). As expected, the highly porous PPY/CNXL nanocomposite film showed an improved capacitance performance (336 F g^{-1}) when compared to that of PPY electrodeposited with chloride anions (258 F g^{-1}) in the same work [17]. Electrochemical impedance spectroscopy (EIS) further demonstrated that the charging/discharging of the PPY/CNXL nanocomposites took place much faster than for pure PPY. Notably, the capacitance of the PPY/CNXL increased significantly compared to PPY in the negative potential region (Fig. 2.4). When PPY is reduced, it becomes non-electron-conductive and the current decreased rapidly from −0.2 V (vs. Ag|AgCl) onwards to more negative potentials. With the incorporation of the relatively large, immobile and negatively charged CNXL into PPY, it was apparent that the negative potential limit at which the composite maintains conductivity was extended to ca. −0.6 V. This is possibly due to the negative charge on the embedded CNXLs electrostatically repelling the electrons on the PPY chains, making electron removal easier and thereby shifting the oxidation potential negative (so films become reduced only at more negative potentials). This phenomenon had been previously reported for PPY/CNT nanocomposites [7] but the effect was partly attributed to the excellent conductivity of embedded CNTs [7, 8]. However, as CNXLs are non-conducting, this suggest that the high current of the PPY/CNXL nanocomposite at the negative potentials cannot be due to the conductivity of the supporting scaffold. When comparing the electrochemical performance between PPY/CNXL and PPY/CNT, it was also found that the PPY/CNXL nanocomposites

Fig. 2.4 CVs obtained at an electrodeposited PPY (*dashed line*) and PPY/CNXL (*solid line*) in 0.1 M KCl at a scan rate of 0.25 V s^{-1}. Each film was electrodeposited with a charge of 0.1 C/cm^2. Reprinted with permission from Ref. [17]. Copyright (2015) American Chemical Society

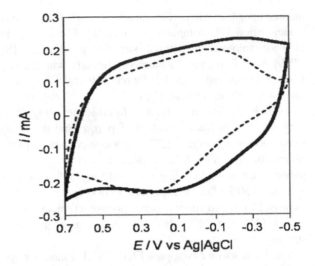

were at least equally capacitive and durable (measured over 5000 potential charge/discharge cycles) as the PPY/CNT nanocomposites, while both nanocomposites performed significantly better than the PPY. This showed, for the first time, that highly conductive additives or support materials such as CNTs are not necessary for the formation of high-performance ECP nanocomposites.

As CNXLs are very strong materials, it was anticipated that their use would also facilitate the growth, by co-electrodeposition, of thicker PPY/CNXL nanocomposite films with a consistent porous structure. This is especially important to make a high total capacitance nanocomposite that will allow effective utilization of the electrode geometric area. It is thus a requirement to develop PPY/CNXL composites to the level where it can be considered for practical supercapacitor applications. The capacitance of the electrode normalized to its geometrical area is called the electrode capacitance (C_E/F cm^{-2}), and this expresses the amount of charge that can be stored per unit area of electrode surface. The more commonly used measure throughout this research field is however the mass-specific capacitance (C_M/F g^{-1}). Typically, high C_M values are often observed only for thin supercapacitor electrode material films (low C_E). When thicker materials are used, the charging/discharging kinetics can become hindered, and as a result the C_M decreases with increasing film thickness. If C_M can be maintained constant with the increasing film thickness, the increase of C_E will be linear with increasing film thickness. Unfortunately, as it is commonly seen that C_M decreases with increasing film thickness, the increase in C_E will also be less than linear with increasing film thickness. At a certain film thickness, C_E will even seize to increase with the addition of more capacitive material, i.e., the electrode reaches its capacitance limit. If C_E approaches its plateau at a relatively low level, the overall capacitance of the electrode will be low and the electrode material may not be suitable for scaling up to commercial and/or practical applications [7, 8, 19]. The challenge is therefore to achieve both a high C_E and C_M values in the new electrode material. This has been achieved with the PPY/CNT nanocomposites, where C_E values of 2.35 F cm^{-2} have been reached by electrodeposition [7], compared to only 0.7 F cm^{-2} for PPY electrodeposited with chloride anions and without supporting additives [8]. It was found that for PPY/CNXL films, the C_E increased linearly with the deposition charge up to 1.54 F cm^{-2} (measured with EIS) and 2 F cm^{-2} (measured with the galvanostatic method), with an average C_M of 240 F g^{-1} [20]. The thick PPY/CNXL film was found to be porous throughout its full thickness (Fig. 2.5), meaning that CNXLs are very suitable for the fabrication of porous and thick, and therefore high total and mass specific capacitance, PPY nanocomposites. A prototype supercapacitor made of two of thick PPY/CNXL electrodes (a symmetric supercapacitor), when subjected to a potential cycling test, lost only 30% of capacitance after 10,000 cycles and about 50% after an excessive 50,000 cycles [20]. Putting this into perspective, as most ECP based super-capacitors have a typical cycle-life of only a few thousand cycles [21], the PPY/CNXL supercapacitor are considered to have excellent stability.

Further work on developing ECP/CNXL nanocomposites by co-electrodeposition has found that CNXLs were also suitable supporting additives for the co-electrodeposition of PANI/CNXL nanocomposite films [22]. The mechanisms by which the CNXLs provide support to the PANI/CNXL nanocomposites were similar to its support in PPY/CNXL nanocomposites. When a thick PANI/CNXL nanocomposite film was compared with a PANI film electrodeposited with chloride anions without supporting additive (Fig. 2.6), it was found that the PANI/CNXL nanocomposite film maintained a capacitive response even at a fast scan rate (Fig. 2.6b). The PANI film on the other hand did not respond to the fast charging and instead showed a resistive response [22]. The EIS Nyquist plot (Fig. 2.6c) of the PANI/CNXL film was characteristic of a straight open pore structure with high accessibility for the electrolyte. The PANI film showed characteristics of narrow-necked quasi-spherical pores that allowed only limited electrolyte access [22, 23]. It was obvious that the presence of the strong CNXLs embedded within the PANI/CNXL nanocomposite structures gave rise to this difference as their strength

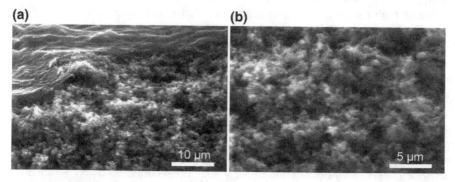

Fig. 2.5 SEM images of a fractured thick PPY/CNXL film cross-section. Reproduced from Ref. [20] with permission from The Royal Society of Chemistry

Fig. 2.6 a and **b** CVs recorded at an electrodeposited PANI/CNXL film (*red lines*) and a PANI film (*black lines*) in 1 M HCl. The scanrates were **a** 0.02 V s^{-1} and **b** 0.25 V s^{-1}. **c** Nyquist plots obtained from EIS of the PANI/CNXL film (*squares*) and the PANI film (*circles*) at 0.6 V in 1 M HCl. The knee frequencies were 0.3 Hz for the PANI film and 5.5 Hz for the PANI/CNXL film. Each film has been electrodeposited to a charge density of 10 C cm^{-2}. Figure from Liew et al. [22] (Color figure online)

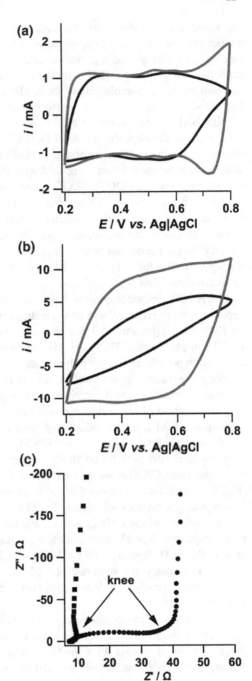

supported the formation of the straight open pore structure even when the nanocomposite films were deposited to a high thickness. The thick PANI/CNXL reached a total electrode capacitance of 2.07 F cm^{-2} with a corresponding specific capacitance of 440 F g^{-1} [22]. Although the thick PANI without the CNXL support can also reach comparable values for its electrode specific capacitance, as shown in Fig. 2.6a, its electrochemical performance is inferior compared to the thick PANI/CNXL, as demonstrated by the fast scan rate CV and the EIS (Fig. 2.6b, c).

The co-electrodeposition of PEDOT/CNXL nanocomposites has also been attempted. When thin films are prepared, the PEDOT/CNXL nanocomposite films also showed improvement when compared with PEDOT films [22]. However, attempts to fabricate thick PEDOT/CNXL nanocomposite films by co-electrodeposition have been unsuccessful. This was because the EDOT monomer was not soluble in water and a mixed water/acetonitrile solvent system was required to prepare the mixture for electrodeposition, and this mixture is unstable for long periods of time, required for the deposition of thick films [22]. Similar effects have also been reported during unsuccessful attempts to fabricate thick PEDOT/CNT nanocomposites by co-electrodeposition [7].

PPY/CNXL nanocomposites have also been fabricated by in situ chemical deposition more recently, and their electrochemical performance has been reported on [24]. The oxidation of Py monomers, dissolved in a 0.2 wt% dispersion of CNXLs in a mixture of HClO$_4$ (1 M) and ethanol (v/v = 1/1), were performed with ammonium persulfate and the polymerization and composite formation was allowed to take place under vigorous stirring for 24 h. The optimal ratio of Py to CNXL in the reaction mixture was determined from both the morphological characterization, i.e., the structure of the nanocomposite product and the coating of the PPY on the individual CNXLs, and conductivity testing. It was found that, under optimum conditions, the PPY coating on the CNXL templates had the highest conjugation length which also manifested in the highest conductivity, more than twice than when uncoated CNXLs were in excess or when unsupported PPY was in excess (Fig. 2.7). This finding shows that the conductivity of PPY can be increased simply by templating it on non-conducting CNXLs. It also shows that the rod-like structure and the surfaces of the CNXLs are suitable for the growth of high conductivity ECP nanocomposites. The electrochemical capacitance of an electrode assembled of the in situ chemical deposited PPY/CNXL was 248 F g^{-1} when characterized using cyclic voltammetry at a scan rate of 0.01 V/s, and lower than the 336 F g^{-1} found by co-electrodeposition by Vix-Guterl et al. [25]. Unfortunately, no stability studies were performed.

Recent work on in situ chemical deposited PPY/CNXL nanocomposites focused on efforts to improve the attachment of PPY to the CNXL surface and to eliminate the formation of excess PPY particles and improve the regularity of the PPY coatings [26]. This is achieved by modifying the CNXL surface with a thin layer of physically adsorbed poly(N-vinyl-pyrrolidone) (PVP). As an amphiphilic species, the adsorbed PVP mediates the surface of the CNXLs from hydrophilic to more hydrophobic, on which PPY favorably templates. In this study, optimization on the

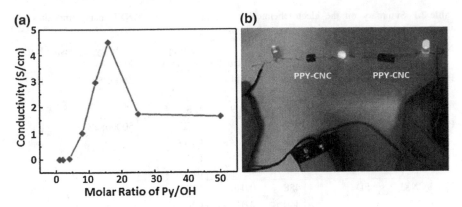

Fig. 2.7 a Optimisation of Py monomer to CNXL amount for highest conductivity, and **b** a demonstration of pressed PPY/CNXL sheets conducting electricity. Figure from Wu et al. [24]

ratio of PVP to CNXLs was carried out and the most suitably modified CNXL surface for PPY templating was based on morphological characterization and conductivity testing. At the optimum, the resulting PPY/CNXL nanocomposite had a conductivity increase of 7 times over that of the PPY/CNXL nanocomposite prepared with CNXLs that were not pre-treated with PVP (36.9 vs. 4.9 S cm^{-1}). It was suggested that this improvement was due to improved attachment and distribution of PPY over the CNXLs, which facilitates the growth of highly structured PPY. However, it is worth mentioning that the pH conditions and the oxidizing agents used for these samples were not kept the same (neutral conditions with FeCl$_3$ oxidizing agents when PVP-treated CNXLs were used, and acidic conditions with ammonium persulfate oxidizing agent when untreated CNXLs were used). Therefore there is a possibility that the neutral conditions and perhaps the use a better oxidizing agent FeCl$_3$ can give rise to the improved performance, but unfortunately this was not investigated by the authors. The electrochemical capacitance of the highly conductive PPY/PVP/CNXL was 322.6 F g^{-1} when characterized using cyclic voltammetry at a scan rate of 0.01 V s^{-1}. A potential cycling stability test was carried out for this nanocomposite and for 1000 cycles only a 8.9% reduction in capacitance was measured compared to a 36.9% loss for the PPY/CNXL nanocomposite made with CNXLs that were not pre-treated with PVP.

These works on the preparation of ECP/CNXL nanocomposites as supercapacitor electrode materials have shown that CNXLs are very suitable templates for the formation of high performance ECP nanocomposites. Both electrodeposition and chemical deposition methods have been employed for the fabrication of these nanocomposites and the results from all these works have been consistent. Through electrodeposition, PPY/CNXL and PANI/CNXL nanocomposites with a high mass specific and total electrode capacitance (above 2 F cm^{-2}) have been fabricated (Table 2.1) [20, 22]. On the other hand, while high mass specific capacitance values

Table 2.1 Summary on the electrochemical performance of ECP/CNXL nanocomposites as supercapacitor electrode materials

Nano-composite	Fabrication method	C_M (F g^{-1})	C_E (F cm^{-2})	Testing condition	Stability, cycles; loss (%)	Ref.
PPY/CNXL	ED	336	0.011	CV (0.25 V s^{-1})	5000; 50	[17]
		256	0.009	EIS		
PPY/CNXL	ED	240	2	GCD	50000; 53	[20]
		140	1.2	GCD		
		70	0.6	CV (0.1 V s^{-1})		
PANI/CNXL	ED	488	0.046	EIS	5000; 51	[22]
		440	2.07	CV (0.02 V s^{-1})		
PPY/CNXL	CD	248	0.070	CV (0.01 V s^{-1})		[24]
PPY/CNXL	CD	322.6	0.091	CV (0.01 V s^{-1})	1000; 9	[26]

ED Electrodeposition, *CD* Chemical deposition, C_M Mass specific capacitance, C_E Total electrode capacitance, *CV* Cyclic voltammetry, *EIS* Electrochemical impedance spectroscopy and *GCD* Galvanostatic charge discharge

were also reported for the chemically deposited PPY/CNXL nanocomposites, the electrochemical characterization was performed with only a small amount of the material used (20 µg) [24, 26]. Back calculation shows that the corresponding total electrode capacitance in those cases were rather small (0.07 and 0.09 F cm^{-2}, compared to 2 F cm^{-2} demonstrated for the electrodeposited version as shown in Table 2.2). In other words, the high total electrode capacitance has not yet been demonstrated for the chemically deposited PPY/CNXL nanocomposites. Given the obvious advantages of the chemical deposition method for producing PPY/CNXL nanocomposites at larger quantities more readily, the characterization of an electrode made with a considerable quantity of the material should be carried out in the near future. It is also important to point out that for large-scale production, both chemical deposition and co-electrodeposition of PPY/CNXL can be used. Although it is true that in most cases chemical deposition is easier to scale-up, electrodeposition processes are also common on an industrial scale e.g., in electroplating of metal surfaces. These large systems could be easily re-engineered for the electrodeposition of ECP nanocomposites on electrode surfaces. Thus far the work on ECP/CNXL nanocomposites has proven the concept that the use of CNXLs can in many ways improve the ECP nanocomposite performance. More importantly, they also show that a conducting support such as CNTs is not required, as previously believed, for the making of high performance ECP nanocomposites.

Table 2.2 Highlights of some of the PPY/CNF based electrode for supercapacitors

Nanocomposite	Highlights	Ref.
PPY/CNF	Symmetrical supercapacitor assembled, device capacitance was 32.4 F g^{-1} and stable over 4000 cycles at 0.5 A g^{-1}	[27]
PPY/CNF reinforced with carbon fiber	High conductivity ternary composite, fast charge/discharging capabilities. Lower mass specific capacitance due to inclusion of micron sized carbon fiber, around 93 F g^{-1}	[28]
PPY/CNF	CNF of bacterial origin, reaction conditions optimized to achieve high conductivity of 77 S cm^{-1}. Specific capacitance was 316 F g^{-1}	[29]
PPY/CNF membrane	PPY deposited onto a CNF membrane. Specific capacitance was 459.5 F g^{-1}	[30]
PPY/CNF	High capacitance device, up to 15.2 F, corresponding to 38.3 F g^{-1} for the total electrode mass and 2.1 F cm^{-2} normalized to the electrode area. Can charge up to 95% of capacity within 22 s using a potential step charge method. More than 80% initial capacitance retained after 10,000 cycles	[31]
PPY/CNF compressed	Compact electrode reduces the dead volume and dead weight in supercapacitor assembly. Very high electrode capacitance of 5.66 F cm^{-2} and volumetric capacitance of 236 F cm^{-3}	[32]

2.3 ECP/CNF Nanocomposites

Aside from CNXLs, cellulose nanofibrils (CNF) have also been used in the preparation of ECP nanocomposites. Compared to CNXLs, CNFs are longer and more flexible as they contain some amorphous sections in between crystalline parts [33]. As a result of their higher flexibility, CNFs forms flexible networks whereas the rigid-rod CNXL networks are more brittle. The CNF network flexibility is very useful for the fabrication of high performance flexible electrodes and supercapacitors, which has recently drawn a lot of research interest. Incidentally, research efforts into developing high performance flexible supercapacitors also frequently used bulk cellulose materials (paper [34–39], microfibers [40], and textile cotton [41, 42]) as the flexible substrates on which electrode materials were deposited. More importantly, there have also been reports on the development of flexible supercapacitors on CNF-based fabric materials [43–46]. In addition, strong aerogel materials made by crosslinking CNF networks used for the deposition of a thin-film supercapacitor and battery system have also been reported recently [47, 48]. However, since the focus of this section is on the ECP/CNF nanocomposites, we will not further discuss on the use of bulk cellulose or CNF-based fabric materials as the underlying substrate.

PPY/CNF nanocomposites have been fabricated using the in situ chemical deposition method [49]. For the making of this PPY/CNF nanocomposite, the Py monomers were mixed with CNF in water and the polymerization of Py was performed subsequently with FeCl$_3$ as the oxidizing agent. A polymerization time of 15 min was employed and coating of PPY onto the CNF took place in this short time. The nanocomposite was then collected on a filter paper by filtration. It was found that, after drying under atmospheric conditions, the PPY/CNF nanocomposite structure did not collapse and maintained a porous gel structure. This highly porous PPY/CNF nanocomposite had a high surface area of 90 $m^2 g^{-1}$ and a conductivity of 1.5 S cm^{-1}. Although the conductivity of 1.5 S cm^{-1} appears to be low compared to those of the chemically deposited PPY/CNXL nanocomposites (e.g., 36.9 S cm^{-1}), the high value for the PPY/CNXL was measured using a compressed sample whereas the PPY/CNF nanocomposite was a self-supporting porous gel [26]. The PPY/CNF nanocomposite had a charge capacity of 289 C g^{-1} (roughly 200 F g^{-1} given the 1.5 V active potential window of the nanocomposite), measured using the CV method. Conductive aerogel composites based on PPY/CNF were subsequently developed [50]. The PPY/CNF nanocomposite aerogel was fabricated by in situ chemical deposition, similar to the procedures described above except the final drying step. Subsequently, by subjecting the wet samples to different drying conditions and methods, aerogels of varying structural characteristics were produced (Figs. 2.8 and 2.9). The best performing PPY/CNF aerogel sample was found to be the supercritical CO_2 dried aerogel (Comp_CO_2 as in Figs. 2.8 and 2.9), due to its higher surface area as supercritical CO_2 drying allows solvent removal without the collapse of the pore structure. The specific charge storage capacity of the supercritically dried PPY/CNF aerogel was 220 C g^{-1} (corresponding to about 200 F g^{-1}, given the 1.1 V active potential window).

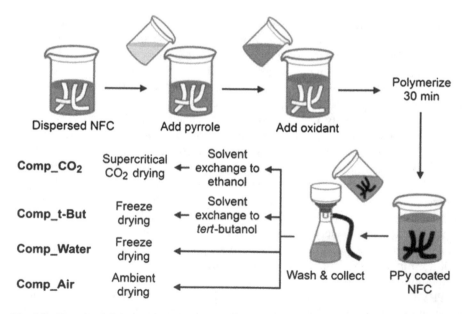

Fig. 2.8 Overview of the synthesis and drying procedures of the PPY/CNF nanocomposites. The wet PPY/CNF nanocomposites were divided to four parts for drying with different procedures. Figure from Carlsson et al. [50]

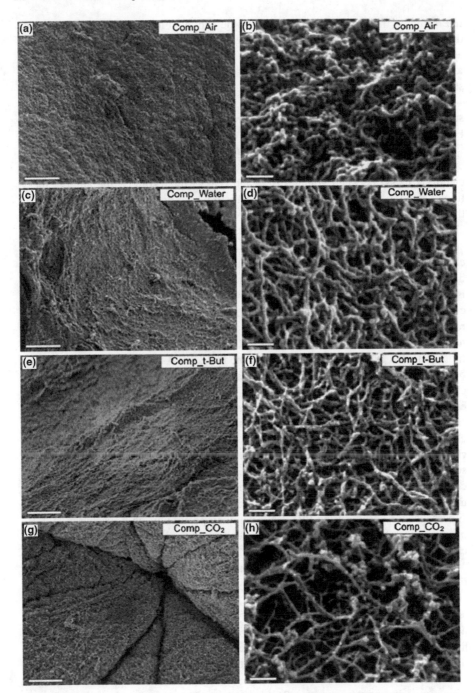

Fig. 2.9 SEM micrographs of PPY/CNF samples prepared using different drying methods as shown in Fig. 2.8. The scale cars of the *left* and *right* panels are 5 μm and 200 nm, respectively. Figure from Carlsson et al. [50]

Interestingly, this is not higher than the value achieved for the nanocomposite film dried under ambient conditions. However, the high porosity of the PPY/CNF aerogel gave rise to significantly faster charge/discharge characteristics, compared to the denser and less porous ambient dried PPY/CNF film. For neither of these studies, stability tests were carried out.

Two PPY/CNF nanocomposite electrodes have also been used in a symmetric supercapacitor assembly [27]. The supercapacitor device capacitance was 32.4 F g^{-1} (total capacitance divided by total electrode mass, corresponding to 129.6 F g^{-1} for each electrode) and had excellent stability over 4000 charge discharge cycles, characterized at 10 mA (corresponding to 0.5 A g^{-1} for 20 mg of the total electrode mass) between voltages of 0 and 0.8 V. Interestingly, the authors mentioned some form of self-protective mechanism which prevented the degradation of the PPY/CNF electrodes when the higher voltage cut-off for the constant current charging/discharging was increased from 0.8 to 1.2 and 1.8 V. When the symmetric supercapacitor (both electrodes PPY as electrochemically active material) is fully charged at roughly 1 V, the PPY material on the positive electrode is in the oxidized state and highly conductive, while the PPY material on the negative electrode is in the reduced form, neutral and resistive. In order to continue delivering a constant current into this system, a higher voltage is required. The extra voltage required will be distributed unevenly over the system and will mostly be sent to the negative electrode because it is significantly more resistive, and therefore requires more potential to drive a charge through the material. As the extra voltage will not be distributed to the positive electrode, the positive electrode will not be further charged. This is beneficial because any extra charge that goes in the positive electrode material will damage it. It was found for the PPY/CNF symmetric supercapacitor that the voltage of the system increased rapidly beyond the fully charged state of 1 V to the higher voltage cut-off of 1.2 and 1.8 V, when the extra charge passed in the system was only very small, therefore giving rise to the so-called self-protective mechanism. This is as far as we know unique for supercapacitors based on ECP supported by non-conducting substrates. The authors of the paper did not mention however, whether this means that the symmetric supercapacitor based on the PPY/CNF nanocomposite will only discharge at 1 V, because the 'self-protective system' implies that there is practically no accessible charge at the voltage above 1 V.

Reinforcing electrodes of PPY/CNF nanocomposites with micron size carbon fibers was found to give rise to a ternary composite with a significantly higher charge/discharge performance than PPY/CNF on its own [28]. The structure of the ternary composite is composed of a highly conductive, rigid micron-scale porous network made up of carbon fibers, which holds the PPY/CNF nanostructures within the micron-scale pores. This highly conducting network of carbon fibers is effectively an extension from the current collector to the whole volume of the composite, and therefore the contact resistance between the otherwise unsupported PPY/CNF nanocomposites to the current collector was significantly reduced. As a result, the charging and discharging of this electrode can take place at a much higher rate and is only limited by the electrolyte transport, than can be achieved for the PPY/CNF nanocomposite without the added carbon fibers. Unfortunately, the addition of large supporting species to a nanocomposite also has its disadvantages, manifested in the overall lower mass specific capacitance. The authors of the paper reported for the ternary composites a charge capacity of 200 C g^{-1} (roughly 200 F g^{-1}) but based only on the mass of PPY (which is 6 parts out of 13 and 6 parts out of 17 for two samples of the ternary composite). This means the overall mass specific capacitance calculated with the total composite mass is between 70 and 93 F g^{-1}, which is lower than the PPY/CNF composite without the carbon support.

Highly conductive PPY/CNF nanocomposites have been fabricated by in situ chemical deposition of PPY on CNF of bacterial origins after optimization of the deposition strategy (Fig. 2.10) [29]. Optimization was carried out on six variables, i.e., the ratio of CNF to PPY, the ratio of Py monomers to FeCl$_3$ oxidants, the acidity of the reaction mixture, the solvent composition (mixture of dimethylformamide and water), the temperature, and time. A very high conductivity of 77 S cm^{-1} was measured on the optimal PPY/CNF nanocomposite using a sample pressed at

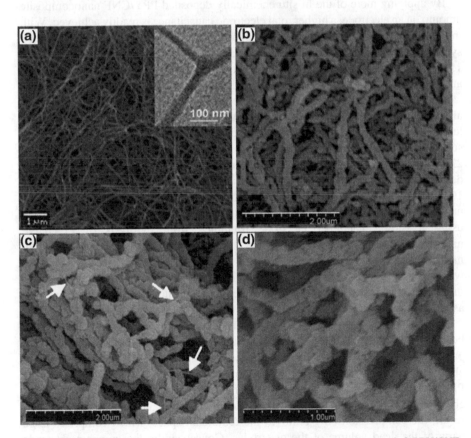

Fig. 2.10 SEM images of **a** Bacterial CNF (inset is a TEM image), **b, c** and **d** PPY/CNF nanocomposites obtained from the optimized reaction protocol. Uncoated CNFs are *arrow-pointed*. Figure from Wang et al. [29]

15,000 Pa. A specific capacitance of 316 F g^{-1} was reported for a 5 mg nanocomposite sample when characterized galvanostatically at 0.2 A g^{-1}. A similar strategy has also been applied to prepare nanocomposites of PANI/CNF using bacterial CNF [51]. The optimized PANI/CNF nanocomposite had a specific capacitance of 273 F g^{-1} equally determined galvanostatically at 0.2 A g^{-1}. This is slightly lower than most PANI-based electrode materials. A possible explanation for this is that the optimization strategy was carried out to give the nanocomposite with the highest conductivity rather than for the highest capacitance. It appears that the focus on conductivity for optimization of capacitance is not suitable. By chemically depositing PPY onto a porous membrane of bacterial CNF, flexible supercapacitor electrodes with a specific capacitance of 459.5 F g^{-1} have been made [30], surpassing the value obtained for the sample optimized for conductivity. Despite this initial high specific capacitance, the nanocomposite membrane suffers a significant loss in capacitance after only a few charge/discharge cycles and this may indicate that the initial high-capacitance composite structure formed is not stable.

By applying more of the in situ chemically deposited PPY/CNF nanocomposite amount on an electrode, a higher total electrode capacitance is readily achieved. With the use of 400 mg of the PPY/CNF nanocomposite spread over two electrodes in a symmetrical supercapacitor assembly operating up to 0.8 V, a cell capacitance of 15.2 F was reached [31]. This corresponds to 38.3 F g^{-1} for the total electrode mass and 2.1 F cm^{-2} normalized to the area of the current collector. This high electrode capacitance is due to the porous composite structure and thin PPY coatings which allowed rapid electrolyte transport. When the contact resistance to the current collectors was minimized, it was demonstrated that a PPY/CNF nanocomposite based supercapacitor of 12.3 F capacitance can be charged to 95% of full capacity within 22 s by using a potential step charging method. Furthermore, the supercapacitor device retained more than 80% of its initial capacitance after being subjected to 10,000 potential step cycles. This demonstrates the capacity of PPY/CNF nanocomposites to perform at the practical commercial level of performance. By using a carbon nanofiber electrode (obtained from pyrolysis of PPY/CNF nanocomposites, shown in Fig. 2.11) as the negative electrode, and a PPY/CNF nanocomposite as the positive electrode, the operational voltage of the supercapacitor was increased from 0.8 to 1.6 V. The doubling of the operational voltage means that the energy storage density of the asymmetric supercapacitor was almost twice as high as the energy density from a symmetrical PPY/CNF supercapacitor device.

Another very interesting characteristic of the PPY/CNF nanocomposites that has been reported is that electrodes based on this material can be mechanically compacted to more than twice their initial density without a significant change in the electrochemical performance [32]. It was found that the mesoporous structure of the PPY/CNF nanocomposite, which is the most important factor in their electrochemical performance, was largely unaffected by the mechanical compression. As a result, the compression only removes the redundant space, which comprises large parts of the dead volume of the electrodes. Consequently, the compact electrode based on the PPY/CNF was found to have a very high volumetric capacitance, and an electrode capacitance of 5.66 F cm^{-2} and a volumetric capacitance of

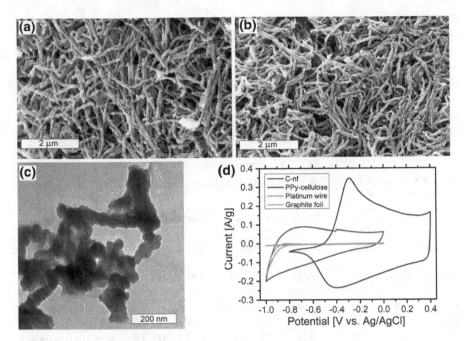

Fig. 2.11 **a** SEM image of a PPY/CNF sample. **b** SEM image of a carbon nanofiber sample derived from pyrolysis of PPY/CNF. **c** TEM image of the carbon nanofiber sample. **d** CVs obtained at a scan rate of 1 mV/s with different working electrodes demonstrating the potential window of the PPY/CNF and carbon nanofiber electrode materials. Figure from Tammela et al. [52]

236 F cm^{-3} could be achieved, which is among the highest ever reported for ECP based electrodes. The retention of the mesoporous structure of the PPY/CNF nanocomposite is possible only because of the strength of the CNF scaffold. The compression loading (from 1 to 5 ton/cm^2) is mainly born by the CNF, as PPY is a mechanically weak material, and the CNF resist a complete pore closure under this compressive load. This is clearly indicated as the porosity of the PPY/CNF nanocomposite decreased from 74% for a sample that is not compressed to 38% after compression at 1 ton/cm^2, while a further compression at 5 ton/cm^2 only resulted in a 12% porosity reduction, giving a final 26% porosity of the fully compressed sample at 5 ton/cm^2. This work also demonstrates the very strong interaction between PPY and the CNF material since the compression has no effect on the individual PPY/CNF nanocomposite fibrous structure. Detachment or deformation of the PPY coating on the individual CNF was not observed. This shows again the excellent compatibility of the nanocellulose materials (both CNF and CNXLs) with ECPs at the nanometer scale.

As can be noticed from this review that all the ECP/CNF nanocomposites discussed in this chapter have been prepared by in situ chemical deposition and as far as we know there is no report on the preparation of ECP/CNF nanocomposite using co-electrodeposition. This is easy to understand as CNF are typically larger and significantly longer than the CNXLs and therefore their transport properties are significantly more sluggish in comparison. CNF are therefore less suitable for electrochemical fabrication (which required fast diffusion of the charge balancing particles). It is thus much more straightforward to alter the PPY behavior in ECP/CNXL composites at the molecular level by controlling structure formation and local coating thickness of the PPY for improved performance. On the other hand, the advantage of using CNF is clearly manifested when it comes to assembling electrodes at the device scale, where the ECP/CNF nanocomposites have

Table 2.3 Waste materials used in the preparation of supercapacitors and their corresponding electrochemical performance parameters

Carbon source	Pyrolysis (°C)	SSA (m²/g)	C_s (F g^{-1})	Electrolyte	Capacitance retention
Potato waste [53]	700	1052	255 at 1 A g^{-1}	2 M KOH	93.7% after 5000 cycles at 5 A g^{-1}
Rice brans [54]	700	2475	265 at 10 A g^{-1}	6 M KOH	87% after 10,000 cycles at 10 A g^{-1}
Coconut shell [55]	800	2440	246 at 0.25 A g^{-1}	0.5 M H$_2$SO$_4$	93% after 2000 cycles at 0.25 A g^{-1}
Corn husks [56]	800	928	356 at 1 A g^{-1}	6 M KOH	95% after 2500 cycles at 5 A g^{-1}
Bamboo [57]	750	169	171 at 1 A g^{-1} 221 at 1 A g^{-1}	1 M KOH 1 M H$_2$SO$_4$	92%/90% after 2000 cycles at 4 A g^{-1}
Fish scale [58]	700	1300	332 at 1 A g^{-1}	6 M KOH	100% after 5000 cycles at 1 A g^{-1}
Hemp [59]	1000	1173	204 at 1 A g^{-1}	1 M LiOH	99% after 10000 cycles at 10 A g^{-1}
Willow catkin [60]	HT, MnO$_2$	234	189 at 1 A g^{-1}	1 M Na$_2$SO$_4$	98.6 after 1000 cycles at 1.0 A g^{-1}
Cabbage [61]	800	3102	336 at 1 A g^{-1}	2 M KOH	95%/after 2000 cycles at 5 A g^{-1}
Soybean curd residue [62]	700	582	215 at 0.5 A g^{-1}	2 M KOH	92% after 5000 cycles at 5 A g^{-1}
Cat tail [63]	850	1951	336 at 2 mV s^{-1}	6 M KOH	Not given
Pomelo peel [64]	600	2105	342 at 1 A g^{-1}	2 M KOH	Not given
Banana peel [65]	1000	1650	206 at 1 A g^{-1}	6 M KOH	98.3 after 1000 cycles at 10 A g^{-1}
Sunflower seed shell [66]	700	2585	311 at 0.25 A g^{-1}	30% KOH	n.g.
Coffee beans [67]	900	1840	361 at 1.0 A g^{-1}	1 M H$_2$SO$_4$	95% after 10,000 cycles at 5 A g^{-1}

shown very promising results both in bulk performance and the ability to create flexible and compressible high-performance nanocomposites. Highlights of some of these works are summarized in Table 2.3.

2.4 Carbonized Polysaccharides

Besides the direct use of polysaccharides in conductive composites as capacitor materials, they can also be used as templates to generate highly porous carbonaceous materials exhibiting high surface area. The main motivation behind this strategy is the renewability of the raw material, the rather low cost and the easy shaping, particularly if soluble derivatives are used. Moreover, the presence of other functional groups such as amines for instance allow for a doping of the final supercapacitor material leading to superior properties. The main experimental parameters to tune the performance of the supercapacitors are therefore the shaping of the polysaccharide (including crosslinking, drying, modification etc.) and the pyrolysis conditions (i.e., atmosphere, temperature, heating ramp, activation). In the following a comprehensive overview is provided covering the literature on available reports on carbonized polysaccharides for supercapacitor application.

One of the most versatile starting materials is paper since it is cheap, and provides a porous network of interconnected individual cellulose fibers. Further papers are mechanically robust and flexible. Therefore, strategies to employ papers in carbonization processes and then to use the resulting freestanding material in supercapacitors have been developed. Particularly filter papers (FP) are very interesting since they do not contain other components than cellulose. Figure 2.12 shows SEM images of FP and its corresponding pyrolysis products obtained at different temperatures [68]. Interestingly, this network sustains upon carbonization whereas the temperature does not seem to influence the morphology at first glance. However, the capacitance determined by galvanostatic charge/discharge and cyclovoltametric experiments, showed a gradual increase at higher temperatures from 0.07 (untreated filter paper) to ca. 120 (carbonized at 1500 °C) F g^{-1} at 5 mV s^{-1}. For the best supercapacitor material obtained at 1500°, a quasi-rectangular shape along the current–potential axis is observed at all scan rates, which reveals a well-defined, electric double-layer capacitive behavior. After 3000 cycles at 1 A g^{-1}, 87% of the capacity was retained which is comparable to other materials such as graphenes and carbon nanotubes.

Recently, Hu et al. extended this approach by addition of another conducting polymer into carbonized paper, namely PANI [69]. PANI was polymerized in situ in the paper substrate and was directly attached to electrodes without any binder or conductive agents. The manufactured materials showed a very high specific capacitance of up to 1090 F g^{-1} at 0.1 A g^{-1} combined with rather low resistance and good cycling stability (84% capacitance at 10 A g^{-1} after 1000 cycles in 1 M H_2SO_4).

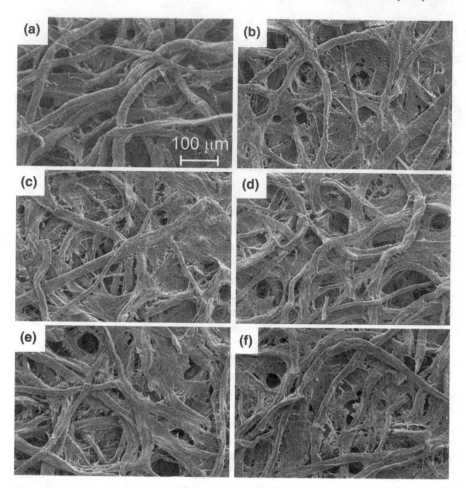

Fig. 2.12 Scanning electron microscopy (SEM) images of filter paper (FP). **a** Without treatment, and with heat-treatment at **b** 600, **c** 1000, **d** 1300, **e** 1500 and **f** 1700 °C [68]. Reproduced with permission from ChemistryOpen, Copyright (2015) John Wiley and Sons

A different approach describes the use of carbonized paper (900 °C) as substrate to grow nanosheets of MnO_2 on the paper surface [70]. This was accomplished by reacting $KMnO_4$ with the carbonized paper in the presence of water yielding nanolayers of MnO_2 with specific capacitance up to 306 F g^{-1} at 0.5 A g^{-1} in 1 M Na_2SO_4 and excellent cycling stability (97% at 100 A g^{-1} after 6000 cycles). Moreover, as for the previous example, any additional binder is not required for the preparation of electrodes, which can largely reduce the interface resistance and ensure high power density.

In a similar manner, MnO_2 nanolayers can also be grown on carbonized fibers for intended use in textiles. Firstly, He et al. used carbonized flax textiles for this purpose which lead to rather low specific capacitance (0.78 F g^{-1} at 0.1 A g^{-1} in

0.1 M Na_2SO_4) due to their relatively low surface area [71]. However, the relaxation time of the system was reported remarkably short (39 ms) allowing for fast charge/discharge cycles (scan rates up to 25 V s^{-1}) and additionally the resistivity is very low. Also the stability of the system was very high and hardly any reduction in efficiency has been observed after 10,000 cycles at 5 A g^{-1}. Nevertheless, the neat carbonized textiles cannot be used in capacitors due to their low specific capacitance but they can serve as excellent substrates for other conducting materials such as MnO_2 nanosheets. In a similar manner as in the previous report such nanosheets were synthesized by exposure of the carbonized textile to aqueous solutions of $KMnO_4$. The best performing materials allowed a current density of 300 A g^{-1}. The specific capacity is 684 F g^{-1} at 2 A g^{-1} and 269 F g^{-1} at 300 A g^{-1} with corresponding energy densities of 47 and 46 W h kg^{-1} while the stability is remarkable (99 and 94.5% respectively after 1000 cycles at 50 A g^{-1}).

Nanofibrous mats are another class of material which provides a fiber network but with much higher surface area than papers. Cai et al. used electrospinning to generate cellulose acetate nanofibers [72]. After deacetylation of the nanofibers, the resulting cellulose was soaked with pyrrole which was in situ polymerized into the nanofiber web, followed by a carbonization step. As a result, N-doped carbon nanofibers with diameters of ca. 600–800 nm were obtained. The procedure is depicted in Fig. 2.13.

At a current density of 0.2 A g^{-1}, the N-doped carbon nanofibers show a significantly higher specific capacitance than the neat carbon nanofibers (236 vs. 105 F g^{-1}). Additionally, the N-doped carbon nanofibers showed a relatively high

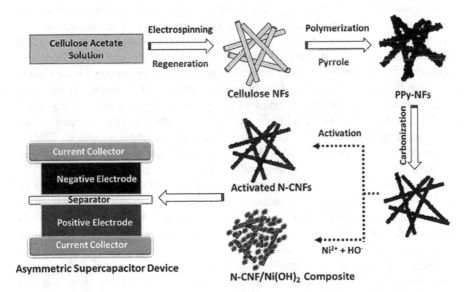

Fig. 2.13 Schematic illustration of a procedure to produce carbonized nanofibers by electrospinning [72]. Reprinted with permission from ACS Applied Materials & Interfaces, Copyright (2015) American Chemical Society

retention of the capacitance when the current density increased (203 F g^{-1} at 1.0 A g^{-1} and 171 F g^{-1} at 10 A g^{-1}) combined with very high stability (98% after 10,000 cycles at 20 A g^{-1}). The authors also demonstrated that this material can be used as negative electrode to build an asymmetric supercapacitor with $Ni(OH)_2$ nanoparticles forming the positive electrode. The nanoparticles were deposited in a solid state reaction directly on the N-doped carbon nanofibers. The resulting composite has a very high specific capacitance of 1045 F g^{-1} at a scan rate of 1 mV s^{-1}. The authors attribute the significant performance increase to a synergistic effect of the $Ni(OH)_2$ nanoparticles with N-doped carbon nanofibers allowing for better interaction with the electrolyte combined with the rather low resistivity of the material. However, at higher scan rate the capacitance decreases since the diffusion of protons in the electrode is kinetically limited leading to partially inaccessibility of the electrode hampering the redox chemistry in the interior of the electrode material.

Another approach to achieve an N-doping of the materials is to add ammonium chloride after the regeneration step of cellulose acetate (see Fig. 2.13) [73]. The addition of ammonium chloride has an additional function in the composite since it stabilizes incompletely regenerated acetate upon heat treatment, increasing the carbon fiber yield. The specific capacity of the carbon nanofibers is increased by the N-doping from 17 to 40 F g^{-1} at a scan rate of 10 mV s^{-1}. However, the rather low capacitance of the materials is probably caused by the rather low specific surface area of the materials in combination with pores in the micron size.

A very similar approach has been reported by Deng et al. [74]. They used blend solutions of cellulose acetate and multiwalled carbon nanotubes (MWNT) and subsequently subjected these to electrospinning. Afterwards, the cellulose acetate was deacetylated and the blend nanofibers were carbonized. It turned out that the activation energy required for the carbonization of cellulose was reduced from 230 to 180 kJ/mol due to the presence of the MWNT. Additionally, the MWNT induced a higher degree of order concomitant with larger crystal sizes inside the nanofibers as proven by XRD studies as well as higher conductivity. The specific capacitance of the resulting material was reported 145 F g^{-1} at a current density of 10 A g^{-1} and 6% loading with MWNT which is significantly higher than the neat carbonized cellulose material (105 F g^{-1}). In a different approach, carbon nanotubes (double and multiwalled) have been added after the regeneration step of the electrospun cellulose acetate mats and subsequently subjected to carbonization [75]. The resulting materials have a specific capacitance of up to 300 F g^{-1} at scan rates of 5 mV s^{-1} which subsequently is reduced to 200 F g^{-1} at 100 mV s^{-1}. However, at higher scan rates (higher than 30 mV s^{-1}) the rectangular shape of the CV curve is not retained. The stability of the materials over 1000 cycles is good and hardly any capacitance is lost.

Besides electrospinning, nanofibers can be obtained by isolation of bacterial celluloses (BC). Such nanofibers have been carbonized and subsequently N-doped using aqueous ammonia solutions by Chen and coworkers [76]. They obtained materials with specific capacitance of up to 200 F g^{-1} at 1.0 A g^{-1} in 2 M H_2SO_4. Moreover, some of the materials feature a remarkable sweep rate behavior and even for scan rates as high as 7 V s^{-1} a rectangular shape of the CV curve was retained. The authors also reported the fabrication and integration of a flexible device by impregnating one of their N-doped carbon nanofibers with a PVA-H$_2$SO$_4$ gel electrolyte which acts as both electrolyte and separator (Fig. 2.14). Interestingly, a loss of capacity upon bending the flat device was not observed and even after 100

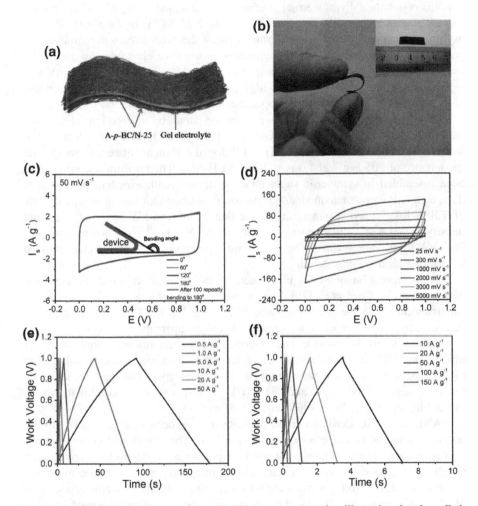

Fig. 2.14 **a** A schematic diagram of the all-solid-state supercapacitor illustrating that the gelled electrolyte can serve as both the electrolyte and separator. **b** A digital photograph of the bent flexible supercapacitor device (2.4 cm × 1.0 cm), showing its good flexibility. **c** CV curves collected at a scan rate of 50 mV s^{-1} for flexible supercapacitor device under different bending angles. Inset is the schematic showing the device under stress and defining the bending angle. **d** Cyclic voltammetry (CV) curves of the flexible supercapacitor at different scan rates. **e** and **f** Galvanostatic charge–discharge curves of the flexible supercapacitor at different current densities [76]. Reprinted with permission from Energy & Environmental Science, Copyright (2013) Royal Society of Chemistry

bending cycles the original capacitance was preserved. The scan rate was varied in a range from 5 mV s^{-1} to 5 V s^{-1}, both exhibiting a rectangular shape of the CV curve. Further, after 5000 charge/discharge cycles, 96% of the initial capacitance has been prevailed.

An asymmetric supercapacitor involving carbonized BC nanofibers was described by Yu et al. [77]. First, they have grown Ni_3S_2 nanoparticles on the carbon nanofibers using a hydrothermal method and obtained materials with specific capacitance as high as 883 F g^{-1} at 2 A g^{-1} in 2 M KOH. In the n next step, an asymmetric supercapacitor was assembled using the Ni_3S_2@carbon nanofibers as positive electrode and the carbon nanofibers as negative electrode materials. The assembled supercapacitor exhibited a specific capacitance of 69 F g^{-1} at 5 mV s^{-1}, high energy density (26 W h kg^{-1}), high power density (425 W kg^{-1}) and rather high stability (97% of the specific capacitance is retained after 2500 cycles).

N,P supercapacitor materials have been prepared by immersion of the BC nanofiber mats in H_3PO_4, $NH_4H_2PO_4$, and H_3BO_3/H_3PO_4 aqueous solutions followed by a carbonization step [78]. The N,P-doped carbon nanofibers show specific capacitance of 205 F g^{-1} at 1.0 A g^{-1} in 2 M H_2SO_4. The resulting materials have been assembled in symmetric supercapacitor devices with excellent performance data, e.g., N,P-carbon nanofibers//N,P-carbon nanofibers exhibits an energy density of 7.8 W h kg^{-1} and a maximum power density of 186 kW kg^{-1}. At a current density of 100 A g^{-1} still an energy density of 1.9 W h kg^{-1} and a power density of 26.7 kW h kg^{-1} are retained. The symmetric supercapacitor further is characterized by a rather high cycling stability.

The incorporation of MnO_2 nanosheets in BC derived carbon nanofibers was demonstrated by Chen et al. [79]. They used this material to assemble a symmetric supercapacitor with specific capacitance of 257 F g^{-1} at 1.0 A g^{-1} in 1 M Na_2SO_4. Additionally, they prepared an asym-metrically supercapacitor using MnO_2@ carbon nanofibers as cathode and N-doped carbon nanofibers (synthesized by immersion of BC in urea solutions) which feature a rather high working voltage (up to 2 V). As a consequence, the energy and power density of the supercapacitor are relatively high (33 W h kg^{-1} and 285 kW kg^{-1}, respectively) in combination with rather high cycling stability (96% after 2000 cycles).

PANI doped BC derived carbon nanofibers have been synthesized by in situ polymerization of aniline after carbonization [80]. These N-doped carbon nanofibers networks serve as support to obtain high performance electrode materials such as activated carbon (AC) and as networks to integrate active electrode materials. The presented AC//MnO_2 asymmetric supercapacitor exhibits a specific capacitance of 113 F g^{-1}, an energy density of 63 W h kg^{-1} at an operation voltage of 2.0 V in 1 M Na_2SO_4.

Another type of cellulosic nanomaterial used for the preparation of supercapacitors are cellulose nanocrystals (CNC). These CNCs are highly crystalline and feature excellent mechanical properties. However, when pure CNCs are pyrolyzed microporous materials are obtained which are less suitable for usage in supercapacitors; therefore hybridization is a straightforward strategy to incorporate mesoporous domains. For instance, CNC/silica composites can be easily prepared and

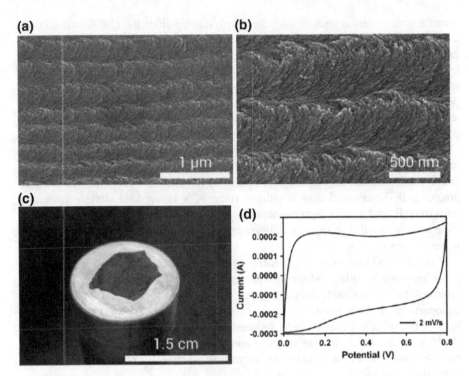

Fig. 2.15 Chiral nematic mesoporous carbon films prepared by pyrolysis of CNC/silica composites exhibit helical twisting morphologies by SEM (**a**, **b**). The freestanding films can be used directly without any binders as electrodes in electrical double-layer supercapacitors (**c**) and display near-ideal capacitor behavior (**d**), with specific capacitances comparable to other state-of-the-art carbon-based supercapacitors [81]. Reprinted with permission from Angewandte Chemie, Copyright (2011) John Wiley and Sons

after carbonization a mesoporous carbonaceous material is obtained which templates the CNC (Fig. 2.15) [81]. The porosity of the material can be adjusted by the amount of silica in the hybrid, ranging from microporous (no silica) to predominantly mesoporous for 35 wt% silica content.

After etching of the silica, the obtained materials are semi-conductive at room temperature and still show nematic phase behavior indicated by the twisting of the left handed CNC. Since the films are freestanding, they can be directly used in supercapacitors without any binding agents. The symmetrical supercapacitor having H_2SO_4 as electrolyte features nearly ideal capacitor behavior. The specific capacitance is 170 F g^{-1} at a current load of 230 mA g^{-1} and decreases at higher current densities. Besides cellulose, also chitin nanocrystals (ChNC) can be used to fabricate supercapacitors and similar to cellulose, additives must be incorporated to provide mesoporosity [82]. In a similar approach, as described above ChNC/silica composites have been prepared and in analogy the pore sizes can be adjusted by the amount of silica in the material. However, compared to CNC derived hybrids, the specific surface area is smaller after the pyrolysis step since the ChNCs are slightly larger than the CNCs. In terms of capacitance, situation is very similar as well exhibiting specific capacitances of 185 F g^{-1} at 230 mA g^{-1}. As for the CNC hybrids, the ChNCs show a gradual loss in capacitance upon higher current densities which is an indication for good energy density of the material paired with low power density. The incorporation of tin oxide nanoparticles into the hybrids led to an improvement in terms of capacitance and energy density and additionally, the power density was significantly improved.

A different approach to realize supercapacitor materials based on cellulose nanocrystals involving metal nanoparticles are depicted in Fig. 2.16. The decisive factor which material is obtained is governed by the used amount of the metal oxide precursor [83]. At small concentrations, route A is favored, at slightly increased ones route B, and when a high concentration is used, materials according to route C have been reported. For supercapacitors, mainly materials prepared by route A and B are of interest.

First, the CNC are crosslinked using cobalt ions resulting in the formation of a gel. Then silica is added and the hybrid is subjected to pyrolysis. Then, the silica is removed by etching and carbon needles are obtained which carry cobalt oxide nanoparticles. It turned out that by increasing the amount of nanoparticles in the hybrid, the specific capacitance is reduced from 91 (case A) to 3 F g^{-1} (case C). Besides the amount, the size of the nanoparticles play a pivotal role, since the particles are ca. 8–10 nm in diameter for case A, whereas for case C the diameter is more than doubled.

Another approach to use CNC in supercapacitors is N-doping. Wu et al. described a surface modification of CNC using a melamine-formaldehyde resin followed by a carbonization step, yielding N-doped carbon nanorods [84]. The resulting materials are hierarchically organized and exhibit pores in the macro, micro and meso regime. Specific capacitance was high as 329 F g^{-1} at 10 mV s^{-1} scan rate and 352 F g^{-1} at 5 A g^{-1} in 1 M sulfuric acid. The materials exhibit rather high cycling stability with only 5% performance loss after 2000 cycles at 20 A g^{-1}.

Templating can also be performed using Mg(OAc)$_2$ 4H$_2$O and Zn(OAc)$_2$ 2H$_2$O and Na-carboxymethyl cellulose (Na-CMC) [85]. Upon carbonization, the inorganic salts decompose and form the corresponding oxides under release of CO$_2$. As a result, highly porous samples are obtained with specific surface areas as high as 1596 m^2 g^{-1}, pore volumes as large as 5.9 cm^3 g^{-1} and hierarchical pore organization. The specific capacitance of the best performing samples was 428 F g^{-1} at 1.0 A g^{-1} paired with high energy density (68.6 W h kg^{-1}). Interestingly, there is hardly any temperature dependence of the material and very similar performance in terms of specific capacitance was determined at 25, 50 and 80 °C respectively.

An interesting way of soft templating was described by Deng and coworkers. They used soda for the generation of CO$_2$ during the carbonization step of different types of mono-(glucose, xylose) and polysaccharides (cellulose, chitin) as well as crude biomass (bamboo, rice straw) creating highly porous carbonaceous materials (specific surface area up to 1893 m^2 g^{-1}) [85]. The best performing materials

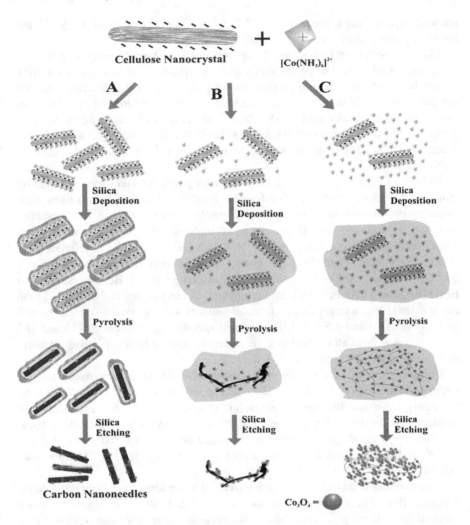

Fig. 2.16 Schematic representation of the synthetic route used to make cellulose-derived amorphous carbon nanoneedles-supported Co_3O_4 nanoparticles (denoted as Co_3O_4/CNN) with different structures and compositions. The synthesis involved: *1* preparation of cellulose nanowhiskers, *2* functionalization of the surfaces of cellulose nanowhiskers with solutions of $[Co(NH_3)_6]^{2+}$ ions with different relative concentrations, *3* deposition of silica shells around the $[Co(NH_3)_6]^{2+}$—modified cellulose nanowhiskers, and *4* pyrolysis of the resulting Co^{2+}/cellulose@SiO_2 core–shell nanowhiskers, and *5* finally etching of the silica shells from the carbonized Co_3O_4/CNN@SiO_2 material. The three different concentrations of $[Co(NH_3)_6]^{2+}$ ions used for the synthesis of the materials led to three different Co_3O_4/CNN materials with different structures and amounts of Co_3O_4, namely: sample Co_3O_4/CNN-A with lower amount of Co_3O_4 compared with carbon, Co_3O_4/CNN-B with comparable amounts of Co_3O_4 and carbon (almost 1:1 ratio by weight), and sample Co_3O_4/CNN-C with larger amount of Co_3O_4 than carbon by weight. Reprinted with permission from RSC Advances, Copyright (2015) Royal Society of Chemistry

showed a specific capacitance of ca. 253 F g^{-1} at a current density of 0.1 A g^{-1} and hardy any capacitance loss after 10,000 cycles.

Materials with even higher specific surface area have been recently described by Wei. They used cellulose, potato starch and eucalyptus wood saw dust as starting materials for the generation of porous carbons [86]. They used an intermediate step for generation of hydrochar in an autoclave at 250 °C before the materials was further processed in a furnace at 700–800 °C. Materials with a specific surface area of up to 2387 m^2 g^{-1} were obtained using this route. The specific capacitance was determined 236 F g^{-1} at a scan rate of 1 mV s^{-1} in TEABF$_4$/acetonitrile (AN) electrolyte.

Hybrids of carbonized cellulose materials with graphene flakes were reported by Huang [87]. Prior to exposure to a furnace, graphene-cellulose hybrid films were manufactured by mixing suspensions of graphene and cellulose. The resulting materials exhibit specific surface areas as high as 1533 m^2 g^{-1} with specific capacitances of 300 F g^{-1} at 5 mV s^{-1} in 6 M KOH with hardly any fading after 5000 cycles and relatively high energy storage performance (67 W h kg^{-1}, 54 W h L^{-1}, and 60 kW kg^{-1} over a 45 s discharge time). Besides KOH, also EMIMBF$_4$ and TEABF$_4$ have been investigated as electrolytes in that study. These electrolytes shows a larger electrochemical window with operation voltages from 0 to 2.8 (TEABF$_4$) and 3.5 V (EMIMBF$_4$) and specific capacitances of 171 and 187 F g^{-1} at 2 A g^{-1}. In all electrolytes, the materials show a high capacitive retention of 97–99% after 5000 cycles at 1 A g^{-1}.

Composites with of cellulose with graphene oxide have been made using a hydrothermal method by Zhang [88]. These composites are characterized by extremely large specific surface areas (up to 3523 m^2 g^{-1}) and excellent bulk conductivity. In ionic liquids (i.e., EMIMBF$_4$, EMIMTFSI, BMIMBF$_4$) these materials show good supercapacitor performance and energy density of 231 F g^{-1} and 98 W h kg^{-1}, respectively, with rather good cycling stability (94–99% after 5000 cycles).

Free standing films of CNT on carbonized cellulose films have been reported by Singsang [89]. The synthesized composites exhibited rather low specific surface areas of ca. 180 m^2 g^{-1}. Nevertheless, the specific capacitance can achieve up to 484 F g^{-1} at a current density of 2 A g^{-1}. The energy density of the CNTs on carbon film synthesized at 600 °C reached 46 W h kg^{-1} at a power density of 0.8 kW kg^{-1} and decreased gradually with increasing power density. The capacity retention is 75% when the current density is increased from 1 to 4 A g^{-1}.

Microcrystalline cellulose was used as starting material for highly porous carbons [90]. In the first step, carbonization at 1050 °C was employed and after cooling the resulting material was subjected to hydrogen peroxide solutions in an autoclave at 250 °C to achieve oxidation of the material. Afterwards the surface groups are removed in another heat treatment step (900°) and the surface area is significantly increased to 927 m^2 g^{-1}. A second oxidation heating cycle even leads to a further increase of the specific surface area (up to 1162 m^2 g^{-1}). The average pore size of the obtained materials was ca. 1 nm which is in the range of the size of the solvated standard organic electrolyte TEABF$_4$ in acetonitrile. The specific capacitance for the material with the highest accessible surface area was reported 100 F g^{-1} in 1 M TEABF$_4$ in acetonitrile.

Cellulose powders have been doped with urea in aqueous NaOH solutions and after carbonization N-doped materials were obtained with high specific surface area (1150 m^2 g^{-1}) and hierarchical pore organization [91]. The N-doped materials show reasonable specific capacitances as high as 177 F g^{-1} at a scan rate of 5 mV s^{-1}. The operation voltage of the supercapacitor was 0.8 V in 1 M H$_2$SO$_4$ and high capacity retention (98%) was observed after 1000 cycles at 20 mV s^{-1}.

Babel reported the use of viscose fibers as carbon source for supercapacitors [92]. Viscose fibers are highly oriented due to the manufacturing process and feature a rather high porosity, exhibiting slit like pores with sizes in the micron range. Upon heat treatment, these pores are altered and, depending on the conditions either micro or mesopores are obtained. One possibility to control this temperature triggered phenomenon is to use resins for impregnation of the fibers whereas the wettability of the fibers with resin can be used to tune the properties of the material. The good fiber wetting resin novolak leads to a homogenous lamellar coatings of the fibers and protects the cellulose fibers against excessive gasification upon heat treatment using elevated temperatures. As a result, micropores are preferentially obtained. In contrast, fibers impregnated with resol generate globular structures on the carbonized fibers leading to a larger extent of mesoporosity in the material. This mesoporosity turned out to be more advantageous in terms of capacity increase (185 F g^{-1} at 2 mV s^{-1}) and anode dynamics when acidic electrolyte systems have been used. For alkaline systems, the use of resins is less straightforward and lower anode capacities have been reported (160 F g^{-1} at 2 mV s^{-1}).

Shrimps shells have been used for the generation N,P doped carbons as well [93]. For this purpose Qu et al. first removed the CaCO$_3$ by HCl treatment followed by immersion in phosphoric acid. After drying the samples were calcined in a temperature range from 400 to 600 °C. The resulting materials show a rather good specific capacitance of 206 F g^{-1} at 0.1 A g^{-1} in 6 M KOH. The doping had a positive impact on the operation window and energy density increased significantly from 2.9 W h kg^{-1} at 0.9 V to 5.2 W h kg^{-1} at 1.1 V, which is close to the decomposition potential of water. The power density accordingly increases from 914 to 1162 W kg^{-1}.

Hydrogels composed of alginates and pectins have been used by Wahid and coworkers to prepare carbonaceous materials [94]. Additionally they incorporated urea in the hydrogel prior to pyrolysis to achieve N-doping of the material and specific surface areas of 837 m^2 g^{-1} have been realized with a high density of mesopores (size ca. 4–5 nm). These materials showed excellent capacitive behavior of 285 F g^{-1} at 1 A g^{-1} with high capacity retention when the current density is increased to 10 A g^{-1} (74%) and 40 A g^{-1} (62%). The cycling stability of the materials is excellent and 96% of the initial capacitance is retained after 2000 charging-discharging cycles at 10 A g^{-1}.

Caragenaan microspheres have been used for supercapacitors by Fan et al. [95]. They subjected caragenaan solutions to hydrothermal treatment in an autoclave to

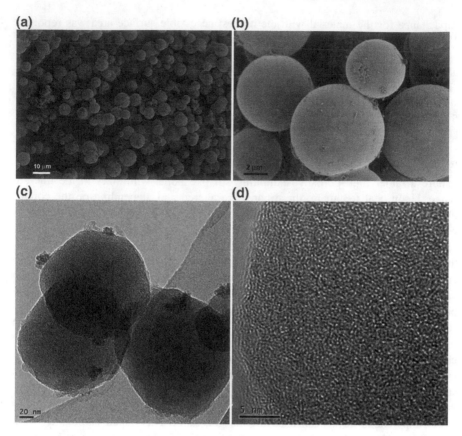

Fig. 2.17 TEM images of caragenaan microspheres before and after carbonization [95]. Reprinted with permission from Journal of Power Sources, Copyright (2014) Elsevier

obtain a kind of hydrochar which is subsequently carbonized in a furnace at elevated temperature. The obtained microspheres (diameter 3–6 μm, Fig. 2.17) exhibit an extremely high specific surface area of up to 2502 $m^2 g^{-1}$ compared to just 14 $m^2 g^{-1}$ of the hydrochar. The specific capacitance of the carbon microspheres reaches 261 F g^{-1} at 0.5 A g^{-1} in 6 M KOH.

Dextrans have been used as precursors for supercapacitors with incorporated iron oxides. For this purpose, iron nitrate was combined with a dextran solution in the presence of ammonia [96]. After carbonization, materials with a specific surface area of ca. 64 $m^2 g^{-1}$ were obtained featuring mesopores with sizes ranging from 3 to 15 nm. Despite the rather low surface area, the materials showed a rather good specific capacitance of 315 F g^{-1} in 2 M KOH solution at a scan rate of 2 mV s^{-1} and good capacity retention (88.9%) after 1500 charge–discharging cycles with energy density of 37 W h kg^{-1}.

Carbon aerogels have been reported by Hao et al. from cellulose isolated from Bargasse, a waste material derived from sugar cane (Fig. 2.18) [97]. They

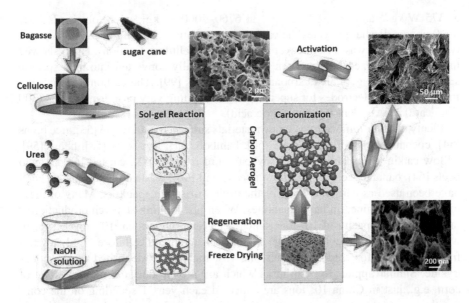

Fig. 2.18 Schematic of the fabrication of highly porous bagasse-derived carbon aerogels with hierarchical pore structure according to Hao et al. [97]. Reprinted with permission from Nanoscale, Copyright (2014) Royal Society of Chemistry

employed a sol gel process to introduce mesoporosity and after drying and subsequent carbonization a carbon aerogel was obtained. Further activation steps allowed to fine tune the porosity and to increase specific surface area up 1892 $m^2 g^{-1}$. The resulting specific capacitance is using a symmetric supercapacitor was about 142.1 $F g^{-1}$ at 0.5 $A g^{-1}$ in 6 M KOH with excellent capacitance retention of 93.86% over 5000 cycles, which translates to the highest energy density of 19.74 $W h kg^{-1}$ and a power density of 0.5 $kW kg^{-1}$.

While all the above mentioned examples use a rather pure form of polysaccharide starting material there are also other approaches which aim at facilitating the production of carbon based materials from biomass waste products either in agriculture or directly from industrial processes. In such waste materials, there is a large variety in terms of composition of polysaccharides but also other materials such as lignins, proteins and inorganics. The main motivation is to save cost for the isolation and separation of lignocellulosic components in such waste materials.

One of such industrial waste materials is paper pulp mill sludge, an ill-defined material which contains besides cellulose many other components [98]. This sludge was hydrothermally converted to hydrochar and subsequently carbonized at high temperatures. By using $TEABF_4$ a maximum capacitance of 166 $F g^{-1}$, and a Ragone curve yielding 30 $W h kg^{-1}$ at 57 $W kg^{-1}$ and 20 $W h kg^{-1}$ at 5450 $W kg^{-1}$. Ionic liquid electrolytes (EMIMTFSI and BMIMTFSI) result in slightly higher capacitances of 180–190 $F g^{-1}$ with up to 62% retention between 2 and 200 mV/s. Energy density-power density couples of 51 $W h kg^{-1}$

at 375 W kg^{-1} and 26–31 W h kg^{-1} at 6760–7000 W kg^{-1} were obtained. After 5000 charge-discharge cycles the capacitance retention is as high at 91%. Another type of industrial waste from paper mills are hemicellulose fractions. Falco showed that such hydrolyzed fractions can be hydrothermally converted into high specific surface (2300 m^2 g^{-1}) area carbonaceous materials [99]. The carbon materials are further tested as electrodes for supercapacitors, yielding very promising results (300 F g^{-1} at 0.25 A g^{-1} in 0.5 M sulfuric acid).

Plenty of other lignocellulosic raw materials such as potato waste [53], rice brans [54], coconut shell [55], corn husk [56], bamboo [57], fish scale [58], hemp [59], willow catkin [60], cabbage [61], soybean curd residue [62], cat tail [63] pomelo peels [64], banana peel [65], sunflower seed shell [66], waste coffee beans [67] and have been used as carbon source for the design of supercapacitors. Many of these materials are obtained in large quantities. When just looking at starch production, a factory with a capacity of 10^6 tons starch per year produces ca. 10^4 tons of potato waste residues, which contains large amounts of starch and celluloses. Only a small fraction is used for feeding animals, the vast majority is deposited on the countryside. Similar applies to corn husks which are a byproduct during harvesting of corn, e.g., just in China 10^5 tons are collected each year. Depending on the composition a variety of carbonaceous materials can be obtained and an overview is given in Table 2.3.

References

1. Lota, K., Khomenko, V., Frackowiak, E.: Capacitance properties of poly (3,4-ethylenedioxythiophene)/carbon nanotubes composites. J. Phys. Chem. Solids **65**, 295 (2004)
2. Wu, M.Q., Snook, G.A., Gupta, V., Shaffer, M., Fray, D.J., Chen, G.Z.: Electrochemical fabrication and capacitance of composite films of carbon nanotubes and polyaniline. J. Mater. Chem. **15**, 2297 (2005)
3. Peng, C., Snook, G.A., Fray, D.J., Shaffer, M.S.P., Chen, G.Z.: Carbon nanotube stabilised emulsions for electrochemical synthesis of porous nanocomposite coatings of poly [3,4-ethylene-dioxythiophene]. Chem. Commun. 4629 (2006)
4. Chen, G.Z., Shaffer, M.S.P., Coleby, D., Dixon, G., Zhou, W.Z., Fray, D.J., Windle, A.H.: Carbon nanotube and polypyrrole composites: coating and doping. Adv. Mater. **12**, 522 (2000)
5. Frackowiak, E., Khomenko, V., Jurewicz, K., Lota, K., Beguin, F.: Supercapacitors based on conducting polymers/nanotubes composites. J. Power Sources **153**, 413 (2006)
6. Khomenko, V., Frackowiak, E., Beguin, F.: Determination of the specific capacitance of conducting polymer/nanotubes composite electrodes using different cell configurations. Electrochim. Acta **50**, 2499 (2005)
7. Peng, C., Jin, J., Chen, G.Z.: A comparative study on electrochemical co-deposition and capacitance of composite films of conducting polymers and carbon nanotubes. Electrochim. Acta **53**, 525 (2007)
8. Hughes, M., Chen, G.Z., Shaffer, M.S.P., Fray, D.J., Windle, A.H.: Electrochemical capacitance of a nanoporous composite of carbon nanotubes and polypyrrole. Chem. Mater. **14**, 1610 (2002)
9. Heath, L., Thielemans, W.: Cellulose nanowhisker aerogels. Green Chem. **12**, 1448 (2010)

10. Tanaka, R., Saito, T., Isogai, A.: Cellulose nanofibrils prepared from softwood cellulose by TEMPO/NaClO/NaClO$_2$ systems in water at pH 4.8 or 6.8. Int. J. Biol. Macromol. **51**, 228 (2012)
11. Kaushik, A., Singh, M., Verma, G.: Green nanocomposites based on thermoplastic starch and steam exploded cellulose nanofibrils from wheat straw. Carbohydr. Polym. **82**, 337 (2010)
12. Eichhorn, S.J., Baillie, C.A., Zafeiropoulos, N., Mwaikambo, L.Y., Ansell, M.P., Dufresne, A., Entwistle, K.M., Herrera-Franco, P.J., Escamilla, G.C., Groom, L., Hughes, M., Hill, C., Rials, T.G., Wild, P.M.: Review: current international research into cellulosic fibres and composites. J. Mater. Sci. **36**, 2107 (2001)
13. Samir, M., Alloin, F., Dufresne, A.: Review of recent research into cellulosic whiskers, their properties and their application in nanocomposite field. Biomacromolecules **6**, 612 (2005)
14. Hubbe, M.A., Rojas, O.J., Lucia, L.A., Sain, M.: Cellulosic nanocomposites: a review. Bioresources **3**, 929 (2008)
15. Eichhorn, S.J.: Cellulose nanowhiskers: promising materials for advanced applications. Soft Matter **7**, 303 (2011)
16. Eichhorn, S.J., Dufresne, A., Aranguren, M., Marcovich, N.E., Capadona, J.R., Rowan, S.J., Weder, C., Thielemans, W., Roman, M., Renneckar, S., Gindl, W., Veigel, S., Keckes, J., Yano, H., Abe, K., Nogi, M., Nakagaito, A.N., Mangalam, A., Simonsen, J., Benight, A.S., Bismarck, A., Berglund, L.A., Peijs, T.: Review: current international research into cellulose nanofibres and nanocomposites. J. Mater. Sci. **45**, 1 (2010)
17. Liew, S.Y., Thielemans, W., Walsh, D.A.: Electrochemical capacitance of nanocomposite polypyrrole/cellulose films. J. Phys. Chem. C **114**, 17926 (2010)
18. Habibi, Y., Chanzy, H., Vignon, M.R.: TEMPO-mediated surface oxidation of cellulose whiskers. Cellulose **13**, 679 (2006)
19. Snook, G.A., Peng, C., Fray, D.J., Chen, G.Z.: Achieving high electrode specific capacitance with materials of low mass specific capacitance: potentiostatically grown thick micro-nanoporous PEDOT films. Electrochem. Commun. **9**, 83 (2007)
20. Liew, S.Y., Walsh, D.A., Thielemans, W.: High total-electrode and mass-specific capacitance cellulose nanocrystal-polypyrrole nanocomposites for supercapacitors. RSC Adv. **3**, 9158 (2013)
21. Snook, G.A., Kao, P., Best, A.S.: Conducting-polymer-based supercapacitor devices and electrodes. J. Power Sources **196**, 1 (2011)
22. Liew, S., Thielemans, W., Walsh, D.: Polyaniline- and poly(ethylenedioxythiophene)-cellulose nanocomposite electrodes for supercapacitors. J Solid State Electrochem. 1 (2014)
23. Macdonald, D.D.: Reflections on the history of electrochemical impedance spectroscopy. Electrochim. Acta **51**, 1376 (2006)
24. Wu, X., Chabot, V.L., Kim, B.K., Yu, A., Berry, R.M., Tam, K.C.: Cost-effective and scalable chemical synthesis of conductive cellulose nanocrystals for high-performance supercapacitors. Electrochim. Acta **138**, 139 (2014)
25. Vix-Guterl, C., Frackowiak, E., Jurewicz, K., Friebe, M., Parmentier, J., Beguin, F.: Electrochemical energy storage in ordered porous carbon materials. Carbon **43**, 1293 (2005)
26. Wu, X., Tang, J., Duan, Y., Yu, A., Berry, R.M., Tam, K.C.: Conductive cellulose nanocrystals with high cycling stability for supercapacitor applications. J. Mater. Chem. A **2**, 19268 (2014)
27. Olsson, H., Nystrom, G., Stromme, M., Sjodin, M., Nyholm, L.: Cycling stability and self-protective properties of a paper-based polypyrrole energy storage device. Electrochem. Commun. **13**, 869 (2011)
28. Razaq, A., Nyholm, L., Sjodin, M., Stromme, M., Mihranyan, A.: Paper-based energy-storage devices comprising carbon fiber-reinforced polypyrrole-cladophora nanocellulose composite electrodes. Adv. Energy Mater. **2**, 445 (2012)
29. Wang, H., Bian, L., Zhou, P., Tang, J., Tang, W.: Core-sheath structured bacterial cellulose/polypyrrole nanocomposites with excellent conductivity as supercapacitors. J. Mater. Chem. A **1**, 578 (2013)

30. Xu, J., Zhu, L.G., Bai, Z.K., Liang, G.J., Liu, L., Fang, D., Xu, W.L.: Conductive polypyrrole-bacterial cellulose nanocomposite membranes as flexible supercapacitor electrode. Org. Electron. **14**, 3331 (2013)
31. Nystrom, G., Stromme, M., Sjodin, M., Nyholm, L.: Rapid potential step charging of paper-based polypyrrole energy storage devices. Electrochim. Acta **70**, 91 (2012)
32. Wang, Z., Tammela, P., Zhang, P., Stromme, M., Nyholm, L.: High areal and volumetric capacity sustainable all-polymer paper-based supercapacitors. J. Mater. Chem. A **2**, 16761 (2014)
33. Frackowiak, E., Beguin, F.: Carbon materials for the electrochemical storage of energy in capacitors. Carbon **39**, 937 (2001)
34. Zhang, X.D., Lin, Z.Y., Chen, B., Sharma, S., Wong, C.P., Zhang, W., Deng, Y.L.: Solid-state, flexible, high strength paper-based supercapacitors. J. Mater. Chem. A **1**, 5835 (2013)
35. Pushparaj, V.L., Shaijumon, M.M., Kumar, A., Murugesan, S., Ci, L., Vajtai, R., Linhardt, R. J., Nalamasu, O., Ajayan, P.M.: Flexible energy storage devices based on nanocomposite paper. Proc. Natl. Acad. Sci. U.S.A. **104**, 13574 (2007)
36. Yuan, L.Y., Yao, B., Hu, B., Huo, K.F., Chen, W., Zhou, J.: Polypyrrole-coated paper for flexible solid-state energy storage. Energy Environ. Sci. **6**, 470 (2013)
37. Yuan, L., Xiao, X., Ding, T., Zhong, J., Zhang, X., Shen, Y., Hu, B., Huang, Y., Zhou, J., Wang, Z.L.: Paper-based supercapacitors for self-powered nanosystems. Angew. Chem. Int. Ed. **51**, 4934 (2012)
38. Nyholm, L., Nystrom, G., Mihranyan, A., Stromme, M.: Toward flexible polymer and paper-based energy storage devices. Adv. Mater. **23**, 3751 (2011)
39. Weng, Z., Su, Y., Wang, D.-W., Li, F., Du, J., Cheng, H.-M.: Graphene-cellulose paper flexible supercapacitors. Adv. Energy Mater. **1**, 917 (2011)
40. Gui, Z., Zhu, H.L., Gillette, E., Han, X.G., Rubloff, G.W., Hu, L.B., Lee, S.B.: Natural cellulose fiber as substrate for supercapacitor. ACS Nano **7**, 6037 (2013)
41. Babu, K.F., Subramanian, S.P.S., Kulandainathan, M.A.: Functionalisation of fabrics with conducting polymer for tuning capacitance and fabrication of supercapacitor. Carbohydr. Polym. **94**, 487 (2013)
42. Zhu, L.G., Wu, L., Sun, Y.Y., Li, M.X., Xu, J., Bai, Z.K., Liang, G.J., Liu, L., Fang, D., Xu, W.L.: Cotton fabrics coated with lignosulfonate-doped polypyrrole for flexible supercapacitor electrodes. RSC Adv. **4**, 6261 (2014)
43. Niu, Q., Gao, K., Shao, Z.: Cellulose nanofiber/single-walled carbon nanotube hybrid non-woven macrofiber mats as novel wearable supercapacitors with excellent stability, tailorability and reliability. Nanoscale **6**, 4083 (2014)
44. Kang, Y.R., Li, Y.L., Hou, F., Wen, Y.Y., Su, D.: Fabrication of electric papers of graphene nanosheet shelled cellulose fibres by dispersion and infiltration as flexible electrodes for energy storage. Nanoscale **4**, 3248 (2012)
45. Kang, Y.J., Chun, S.J., Lee, S.S., Kim, B.Y., Kim, J.H., Chung, H., Lee, S.Y., Kim, W.: All-solid-state flexible supercapacitors fabricated with bacterial nanocellulose papers, carbon nanotubes, and triblock-copolymer ion gels. ACS Nano **6**, 6400 (2012)
46. Wang, X., Gao, K., Shao, Z., Peng, X., Wu, X., Wang, F.: Layer-by-Layer assembled hybrid multilayer thin film electrodes based on transparent cellulose nanofibers paper for flexible supercapacitors applications. J. Power Sources **249**, 148 (2014)
47. Hamedi, M., Karabulut, E., Marais, A., Herland, A., Nyström, G., Wågberg, L.: Nanocellulose aerogels functionalized by rapid layer-by-layer assembly for high charge storage and beyond. Angew. Chem. Int. Ed. **52**, 12038 (2013)
48. Nystrom, G., Marais, A., Karabulut, E., Wagberg, L., Cui, Y., Hamedi, M.M.: Self-assembled three-dimensional and compressible interdigitated thin-film supercapacitors and batteries. Nat. Commun. 6 (2015)
49. Nyström, G., Mihranyan, A., Razaq, A., Lindström, T., Nyholm, L., Strømme, M.: A nanocellulose polypyrrole composite based on microfibrillated cellulose from wood. J. Phys. Chem. B **114**, 4178 (2010)

50. Carlsson, D.O., Nystrom, G., Zhou, Q., Berglund, L.A., Nyholm, L., Stromme, M.: Electroactive nanofibrillated cellulose aerogel composites with tunable structural and electrochemical properties. J. Mater. Chem. **22**, 19014 (2012)

51. Wang, H., Zhu, E., Yang, J., Zhou, P., Sun, D., Tang, W.: Bacterial cellulose nanofiber-supported polyaniline nanocomposites with flake-shaped morphology as supercapacitor electrodes. J. Phys. Chem. C **116**, 13013 (2012)

52. Tammela, P., Wang, Z., Frykstrand, S., Zhang, P., Sintorn, I.-M., Nyholm, L., Stromme, M.: Asymmetric supercapacitors based on carbon nanofibre and polypyrrole/nanocellulose composite electrodes. RSC Adv. **5**, 16405 (2015)

53. Ma, G., Yang, Q., Sun, K., Peng, H., Ran, F., Zhao, X., Lei, Z.: Nitrogen-doped porous carbon derived from biomass waste for high-performance supercapacitor. Bioresour. Technol. **197**, 137 (2015)

54. Hou, J., Cao, C., Ma, X., Idrees, F., Xu, B., Hao, X., Lin, W.: From rice bran to high energy density supercapacitors: a new route to control porous structure of 3D carbon. Sci. Rep. **4**, 7260 (2014)

55. Jain, A., Xu, C., Jayaraman, S., Balasubramanian, R., Lee, J.Y., Srinivasan, M.P.: Mesoporous activated carbons with enhanced porosity by optimal hydrothermal pre-treatment of biomass for supercapacitor applications. Microporous Mesoporous Mater. **218**, 55 (2015)

56. Song, S., Ma, F., Wu, G., Ma, D., Geng, W., Wan, J.: Facile self-templating large scale preparation of biomass-derived 3D hierarchical porous carbon for advanced supercapacitors. J. Mater. Chem. A **3**, 18154 (2015)

57. Chen, H., Liu, D., Shen, Z., Bao, B., Zhao, S., Wu, L.: Functional biomass carbons with hierarchical porous structure for supercapacitor electrode materials. Electrochim. Acta **180**, 241 (2015)

58. Wang, J., Shen, L., Xu, Y., Dou, H., Zhang, X.: Lamellar-structured biomass-derived phosphorus- and nitrogen-co-doped porous carbon for high-performance supercapacitors. New J. Chem. **39**, 9497 (2015)

59. Li, Y., Zhang, Q., Zhang, J., Jin, L., Zhao, X., Xu, T.: A top-down approach for fabricating free-standing bio-carbon supercapacitor electrodes with a hierarchical structure. Sci. Rep. **5**, 14155 (2015)

60. Li, Y., Yu, N., Yan, P., Li, Y., Zhou, X., Chen, S., Wang, G., Wei, T., Fan, Z.: Fabrication of manganese dioxide nanoplates anchoring on biomass-derived cross-linked carbon nanosheets for high-performance asymmetric supercapacitors. J. Power Sources **300**, 309 (2015)

61. Wang, P., Wang, Q., Zhang, G., Jiao, H., Deng, X., Liu, L.: Promising activated carbons derived from cabbage leaves and their application in high-performance supercapacitors electrodes. J. Solid State Electrochem. Ahead of Print (2015)

62. Ma, G., Ran, F., Peng, H., Sun, K., Zhang, Z., Yang, Q., Lei, Z.: Nitrogen-doped porous carbon obtained via one-step carbonizing biowaste soybean curd residue for supercapacitor applications. RSC Adv. **5**, 83129 (2015)

63. Fan, Z., Qi, D., Xiao, Y., Yan, J., Wei, T.: One-step synthesis of biomass-derived porous carbon foam for high performance supercapacitors. Mater. Lett. **101**, 29 (2013)

64. Peng, C., Lang, J., Xu, S., Wang, X.: Oxygen-enriched activated carbons from pomelo peel in high energy density supercapacitors. RSC Adv. **4**, 54662 (2014)

65. Lv, Y., Gan, L., Liu, M., Xiong, W., Xu, Z., Zhu, D., Wright, D.S.: A self-template synthesis of hierarchical porous carbon foams based on banana peel for supercapacitor electrodes. J. Power Sources **209**, 152 (2012)

66. Li, X., Xing, W., Zhuo, S., Zhou, J., Li, F., Qiao, S.-Z., Lu, G.-Q.: Preparation of capacitor's electrode from sunflower seed shell. Bioresour. Technol. **102**, 1118 (2011)

67. Rufford, T.E., Hulicova-Jurcakova, D., Zhu, Z., Lu, G.Q.: Nanoporous carbon electrode from waste coffee beans for high performance supercapacitors. Electrochem. Commun. **10**, 1594 (2008)

68. Jiang, L., Nelson, G.W., Kim, H., Sim, I.N., Han, S.O., Foord, J.S.: Cellulose-derived supercapacitors from the carbonisation of filter paper. ChemistryOpen **4**, 586 (2015)

69. Hu, C., He, S., Jiang, S., Chen, S., Hou, H.: Natural source derived carbon paper supported conducting polymer nanowire arrays for high performance supercapacitors. RSC Adv. **5**, 14441 (2015)
70. He, S., Hu, C., Hou, H., Chen, W.: Ultrathin MnO_2 nanosheets supported on cellulose based carbon papers for high-power supercapacitors. J. Power Sources **246**, 754 (2014)
71. He, S., Chen, W.: Application of biomass-derived flexible carbon cloth coated with MnO_2 nanosheets in supercapacitors. J. Power Sources **294**, 150 (2015)
72. Cai, J., Xiong, H., Cai, J., Niu, H., Li, Z., Du, Y., Cizek, P., Lin, T., Xie, Z.: High-performance supercapacitor electrode materials from cellulose-derived carbon nanofibers. ACS Appl. Mater. Interfaces **7**, 14946 (2015)
73. Kuzmenko, V., Naboka, O., Staaf, H., Haque, M., Goeransson, G., Lundgren, P., Gatenholm, P., Enoksson, P.: Capacitive effects of nitrogen doping on cellulose-derived carbon nanofibers. Mater. Chem. Phys. **160**, 59 (2015)
74. Deng, L., Young, R.J., Kinloch, I.A., Abdelkader, A.M., Holmes, S.M., De, H.-D.R.D.A., Eichhorn, S.J.: Supercapacitance from cellulose and carbon nanotube nanocomposite fibers. ACS Appl. Mater. Interfaces **5**, 9983 (2013)
75. Kuzmenko, V., Naboka, O., Haque, M., Staaf, H., Goeransson, G., Gatenholm, P., Enoksson, P.: Sustainable carbon nanofibers/nanotubes composites from cellulose as electrodes for supercapacitors. Energy (Oxford, U.K.) **90**, 1490 (2015)
76. Chen, L.-F., Huang, Z.-H., Liang, H.-W., Yao, W.-T., Yu, Z.-Y., Yu, S.-H.: Flexible all-solid-state high-power supercapacitor fabricated with nitrogen-doped carbon nanofiber electrode material derived from bacterial cellulose. Energy Environ. Sci. **6**, 3331 (2013)
77. Yu, W., Lin, W., Shao, X., Hu, Z., Li, R., Yuan, D.: High performance supercapacitor based on Ni_3S_2/carbon nanofibers and carbon nanofibers electrodes derived from bacterial cellulose. J. Power Sources **272**, 137 (2014)
78. Chen, L.-F., Huang, Z.-H., Liang, H.-W., Gao, H.-L., Yu, S.-H.: Three-dimensional heteroatom-doped carbon nanofiber networks derived from bacterial cellulose for supercapacitors. Adv. Funct. Mater. **24**, 5104 (2014)
79. Chen, L.-F., Huang, Z.-H., Liang, H.-W., Guan, Q.-F., Yu, S.-H.: Bacterial-cellulose-derived carbon nanofiber@MnO_2 and nitrogen-doped carbon nanofiber electrode materials: an asymmetric supercapacitor with high energy and power density. Adv. Mater. **25**, 4746 (2013)
80. Long, C., Qi, D., Wei, T., Yan, J., Jiang, L., Fan, Z.: Nitrogen-doped carbon networks for high energy density supercapacitors derived from polyaniline coated bacterial cellulose. Adv. Funct. Mater. **24**, 3953 (2014)
81. Shopsowitz, K.E., Hamad, W.Y., MacLachlan, M.J.: Chiral nematic mesoporous carbon derived from nanocrystalline cellulose. Angew. Chem. Int. Ed. **50**, 10991 (2011)
82. Yang, X., Cranston, E.D., Shi, K., Zhitomirsky, I.: Cellulose nanocrystal aerogels as universal 3D lightweight substrates for supercapacitor materials. Adv. Mater. **27**, 6104 (2015)
83. Silva, R., Pereira, G.M., Voiry, D., Chhowalla, M., Asefa, T.: Co_3O_4 nanoparticles/cellulose nanowhiskers-derived amorphous carbon nanoneedles: sustainable materials for supercapacitors and oxygen reduction electrocatalysis. RSC Adv. **5**, 49385 (2015)
84. Wu, X., Shi, Z., Tjandra, R., Cousins, A.J., Sy, S., Yu, A., Berry, R.M., Tam, K.C.: Nitrogen-enriched porous carbon nanorods templated by cellulose nanocrystals as high performance supercapacitor electrodes. J. Mater. Chem. A **3**, 23768 (2015)
85. Deng, J., Xiong, T., Xu, F., Li, M., Han, C., Gong, Y., Wang, H., Wang, Y.: Inspired by bread leavening: one-pot synthesis of hierarchically porous carbon for supercapacitors. Green Chem. **17**, 4053 (2015)
86. Wei, L., Sevilla, M., Fuertes, A.B., Mokaya, R., Yushin, G.: Hydrothermal carbonization of abundant renewable natural organic chemicals for high-performance supercapacitor electrodes. Adv. Energy Mater. **1**, 356 (2011)
87. Huang, J., Wang, J., Wang, C., Zhang, H., Lu, C., Wang, J.: Hierarchical porous graphene carbon-based supercapacitors. Chem. Mater. **27**, 2107 (2015)

88. Zhang, L., Zhang, F., Yang, X., Long, G., Wu, Y., Zhang, T., Leng, K., Huang, Y., Ma, Y., Yu, A., Chen, Y.: Porous 3D graphene-based bulk materials with exceptional high surface area and excellent conductivity for supercapacitors. Sci. Rep. **3**, 1408 (2013)

89. Singsang, W., Panapoy, M., Ksapabutr, B.: Facile one-pot synthesis of freestanding carbon nanotubes on cellulose-derived carbon films for supercapacitor applications: effect of the synthesis temperature. Energy Procedia **56**, 439 (2014)

90. Raymundo-Pinero, E., Gao, Q., Beguin, F.: Carbons for supercapacitors obtained by one-step pressure induced oxidation at low temperature. Carbon **61**, 278 (2013)

91. Yun, Y.S., Shim, J., Tak, Y., Jin, H.-J.: Nitrogen-enriched multimodal porous carbons for supercapacitors, fabricated from inclusion complexes hosted by urea hydrates. RSC Adv. **2**, 4353 (2012)

92. Babel, K., Jurewicz, K.: Electrical capacitance of fibrous carbon composites in supercapacitors. Fuel Process. Technol. **77–78**, 181 (2002)

93. Qu, J., Geng, C., Lv, S., Shao, G., Ma, S., Wu, M.: Nitrogen, oxygen and phosphorus decorated porous carbons derived from shrimp shells for supercapacitors. Electrochim. Acta **176**, 982 (2015)

94. Wahid, M., Parte, G., Fernandes, R., Kothari, D., Ogale, S.: Natural-gel derived, N-doped, ordered and interconnected 1D nanocarbon threads as efficient supercapacitor electrode materials. RSC Adv. **5**, 51382 (2015)

95. Fan, Y., Yang, X., Zhu, B., Liu, P.-F., Lu, H.-T.: Micro-mesoporous carbon spheres derived from Carrageenan as electrode material for supercapacitors. J. Power Sources **268**, 584 (2014)

96. Sethuraman, B., Purushothaman, K.K., Muralidharan, G.: Synthesis of mesh-like Fe_2O_3/C nanocomposite via greener route for high performance supercapacitors. RSC Adv. **4**, 4631 (2014)

97. Hao, P., Zhao, Z., Tian, J., Li, H., Sang, Y., Yu, G., Cai, H., Liu, H., Wong, C.P., Umar, A.: Hierarchical porous carbon aerogel derived from bagasse for high performance supercapacitor electrode. Nanoscale **6**, 12120 (2014)

98. Wang, H., Li, Z., Tak, J.K., Holt, C.M.B., Tan, X., Xu, Z., Amirkhiz, B.S., Harfield, D., Anyia, A., Stephenson, T., Mitlin, D.: Supercapacitors based on carbons with tuned porosity derived from paper pulp mill sludge biowaste. Carbon **57**, 317 (2013)

99. Falco, C., Sieben, J.M., Brun, N., Sevilla, M., van der Mauelen, T., Morallon, E., Cazorla-Amoros, D., Titirici, M.-M.: Hydrothermal carbons from hemicellulose-derived aqueous hydrolysis products as electrode materials for supercapacitors. ChemSusChem **6**, 374 (2013)

Conclusions

Although there are several reports in the use of polysaccharides in batteries, their full potential has not been fully explored yet since on one hand mainly cellulose derivatives have been used and systematic approaches to improve the device performance have not been demonstrated so far. The use of polysaccharide combinations for instance or the addition of other conductive polymers or doping agents may lead to materials which are not even of the same quality as the already used ones but which potentially overcome the current performance limits.

© The Author(s) 2017
S. Yee Liew et al., *Polysaccharide Based Supercapacitors*,
Biobased Polymers, DOI 10.1007/978-3-319-50754-5

Printed in the United States
By Bookmasters